Securing the Digital World

Securing the Digital World: A Comprehensive Guide to Multimedia Security is indispensable reading in today's digital age. With the outbreak of digital range and ever-evolving cyber threats, the demand to protect multimedia data has never been more imperative. This book provides comprehensive research on multimedia information security and bridges the gap between theoretical bases and practical applications.

Authored by leading experts in the area, the book focusses on cryptography, watermarking, steganography, and its advanced security solution while keeping a clear and engaging description that sets this book apart in its capability to make complex concepts accessible and practical, making it an incalculable resource for beginners and seasoned professionals alike.

Key Features:

- Detailed study of encryption techniques, including encryption and decryption methods adapted to multimedia data.
- A comprehensive discussion of techniques for embedding and detecting hidden information in digital media.
- A survey of the latest advances in multimedia security, including quantum cryptography and blockchain applications.
- Real-world case studies and illustrations that demonstrate the application of multimedia information security techniques in various initiatives.
- Contributions from computer science and information technology experts offer a comprehensive perspective on multimedia security.

This book is an invaluable help for cybersecurity professionals, IT professionals, and computer and information technology students. *Securing the Digital World* equips readers with the information and tools required to safeguard multimedia content in a cyber-spatiality full of security challenges.

Securing the Digital World

A Comprehensive Guide to Multimedia Security

Edited by

Subhrajyoti Deb and Aditya Kumar Sahu

CRC Press is an imprint of the
Taylor & Francis Group, an **informa** business

Designed cover image: © Shutterstock

First edition published 2025
by CRC Press
2385 NW Executive Center Drive, Suite 320, Boca Raton FL 33431

and by CRC Press
4 Park Square, Milton Park, Abingdon, Oxon, OX14 4RN

CRC Press is an imprint of Taylor & Francis Group, LLC

ISBN: 9781032663623 (hbk)
ISBN: 9781032663630 (pbk)
ISBN: 9781032663647 (ebk)

DOI: 10.1201/9781032663647

Typeset in Sabon
by KnowledgeWorks Global Ltd.

Contents

About the Editors

Dr Subhrajyoti Deb is Assistant Professor in the Department of CSE at ICFAI University Tripura, India. Before joining ICFAI University, he was a postdoctoral visiting scientist in the applied statistics unit at ISI Kolkata, India. He received his M.Tech. and Ph.D. degrees in Engineering from the NIT Agartala and NEHU Shillong, respectively. Dr. Deb is a member of IEEE and CRSI. His research interests include cryptography, information security, data hiding steganography, and IoT. Dr. Deb has one patent filed and has published 30 publications in refereed journals, book chapters, and conference proceedings. He serves as the organizing chair for MICA 2023. He has participated in international conferences as Organizer and Session Chair. He is an editorial board member and a reviewer of SCI-indexed journals. Dr. Deb has previously served as a coeditor for the *Special Issue on Recent Advances in IoT and Its Applications* in the IEEE MMTC Communication Frontier.

Dr Aditya Kumar Sahu is Associate Professor at Amrita School of Computing, Amaravati. Andhra Pradesh, India. Dr. Sahu has been listed in the top 2% cited researchers list two times. He has over 17 years of teaching and research experience. He is working in the research areas of Multimedia Forensics, Digital Image Watermarking, Image Tamper Detection and Localization, Image Steganography, Reversible Data Hiding, and Convolution Neural Network based Data hiding. He completed his Ph.D. degree in Digital Image Steganography and Steganalysis. He is a member of several editorial boards and associate editor in reputed journals.

Contributors

Abdul Rahman Tosho Abdulahi
Department of Computer Science, Institute of Information and Communication Technology, Kwara State Polytechnic
Ilorin, Nigeria

Oluwakemi Christiana Abikoye
Department of Computer Science, University of Ilorin
Ilorin, Nigeria

Aditya Arsh
Department of CSE, NIT Agartala
Jirania, Tripura, India

Akinbowale Nathaniel Babatunde
Department of Computer Science, Kwara State University
Malete, Kwara State, Nigeria

Abhijit Bandyopadhyay
Raniganj Institute of Computer and Information Science
Asansol, West Bengal, India

Debapriya Banik
Department of CSE, ICFAI University Tripura
Mohanpur, Tripura, India

Bubu Bhuyan
Department of Information Technology, North Eastern Hill University
Shillong, Meghalaya, India

Priyanka Biswas
Department of CSE, NIT Agartala
Jirania, Tripura, India

Panem Charanarur
Department of Cyber Security and Digital Forensics, National Forensic Sciences University
Tripura Campus, Tripura, India

Ritam Chatterjee
Department of CIS, Raiganj University
Raiganj, West Bengal, India

Nilanjan Das
Assistant Professor, Siliguri Institute of Technology
Siliguri, West Bengal, India

Sayan Das
Department of CSE, NIT Agartala
Jirania, Tripura, India

Subhrajyoti Deb
Department of CSE, ICFAI University Tripura
Mohanpur, Tripura, India

Srinivasa Rao Gundu
Department of Digital Forensics, School of Sciences, Malla Reddy University
Hyderabad, Telangana, India

Yagnyasenee Sen Gupta
Department of CSE, NIT Silchar
Silchar, Assam, India
and
Department of CSE, ICFAI University Tripura
Mohanpur, Tripura, India

Rupa Jamatia
Department of Information Technology, North Eastern Hill University
Shillong, Meghalaya, India

Nirmalya Kar
Department of CSE, NIT Agartala
Jirania, Tripura, India

Ezekiel Adebayo Ogundepo
African Institute for Mathematical Sciences
Cape Town, South Africa

Roseline Oluwaseun Ogundokun
Department of Multimedia Engineering, Kaunas University of Technology
Kaunas, Lithuania
Department of Computer Science, Landmark University
Omu Aran, Nigeria

Swarna Panthi
Department of Information Technology, School of Technology North-Eastern Hill University
Shillong, Meghalaya, India

P. K. Paul
Department of CIS, Raiganj University
Raiganj, West Bengal, India

R. Saavedra
Azteca University
Chalco, Mexico

Aditya Kumar Sahu
Amrita School of Computing, Amrita Vishwa Vidyapeetham
Amaravati, Andhra Pradesh, India

Monalisa Sahu
Amrita School of Computing, Amrita Vishwa Vidyapeetham
Amaravati, Andhra Pradesh, India

Sushil K. Sharma
Texas A&M University
Texarkana, TX

Chapter 1

Multimedia tamper detection and localization using digital watermarking

Aditya Arsh, Priyanka Biswas, Nirmalya Kar, Debapriya Banik, and Subhrajyoti Deb

1.1 INTRODUCTION

Globally, digital communication advancements have recently experienced exponential expansion. The Internet makes a vast amount of multimedia material easily accessible to everyone on a daily basis. The use and dissemination of digital information have significantly increased as a result. The involvement of sharing and conveying personal information has also been sparked by advancements in technological devices. Broadcasting a digital picture with some information included inside of it is often pretty usual. However, with the development of powerful image processing software and indeed the accessibility of digital photos, it has never been simpler to alter or change an image's contents. Multimedia data stored in digital format may readily be altered or changed, whether on purpose or by the gaffe, applying a number of imaging techniques. To put that in perspective, we don't know whether the content we acquire via the web is authentic or not. As a result, emphasis has been placed on finding an ideal approach capable of successfully verifying the integrity and authenticity of any digital resource.

Watermarking technologies may demonstrate the ownership, copyright, and validity of any digital material [1]. Approaches based on watermarking not only identify but rather accurately pinpoint changed elements. Digital watermarking, an approach for adding an e-signature or a watermark onto a photograph, can secure the perceptual quality, purity, and legality of digital files. A digital watermark method hides the watermark bits inside the image, allowing the receiver to rapidly extract the information and authenticate the image's validity [2,3]. Additionally, employing various digital watermarking techniques can accomplish desirable qualities such as tamper detection as well as localization of tampered areas.

The detection of digital media manipulation has long been an intriguing subject. The importance of digital media has increased with the expansion of its use online. The volume of provided data has significantly risen, especially for images and videos [4,5]. This has unavoidably garnered attention from a wide range of people, including, unhappily, dishonest individuals who would change the supplied content to suit their needs. So, it is essential to

DOI: 10.1201/9781032663647-1

spot tampering before returning to the original image. While a recent study has also incorporated photo recovery, most of the research yet has been on tamper detection [6].

For various reasons, several digital watermarking techniques have been reported during the prior decade. The construction of an image tamper detection and recovery system based on the discrete wavelet transform (DWT) approach is described in Refs. [7,8], where some information has been recovered as the picture's eigenvalue and is encoded in the frequency domain's middle-frequency band. The localization and detection of tampering both have been accomplished with such embedding. According to Nazari et al. [9], a fragile watermarking technique for chaotic map-based digital photo tamper detection has been developed. This method creates watermarks from picture blocks and incorporates them into the image.

For the purpose of detecting tampering with greyscale pictures, Trivedy and Pal [10] have developed a delicate watermarking technique. Using a random sequence, this technique generates a watermark and key matrix. Key matrices are used to embed. To properly locate and identify the changed areas from the watermarked picture, Sahu [11,12] suggests a fragile watermarking approach based on logistic maps. The suggested approach generates the watermark bits using the susceptibility characteristic of the logistic map. The initial intermediate significant bits (ISBs) are logically XORed with the watermark bits to create the rightmost least significant bits (LSBs).

Researchers [13] have developed a delicate watermarking technique based on hashes to identify modified portions of digital images and to protect their integrity and veracity. The greyscale image is originally divided into four sections that do not overlap. Following a discrete cosine transform (DCT) transformation, the DC coefficient of each of these blocks is determined. The DC coefficients are subjected to a hash operation to produce a 16-bit watermark. An Arnold transform is used to jumble the picture blocks and include a 16-bit hash code. Some other studies regarding digital watermarking techniques for media tamper detection and localization are discussed in Table 1.1.

According to the existing studies, we can conclude that several algorithms have been created to retain imperceptibility and payload while offering security. Yet, it has been observed that security breaches are on the increase, raising the issue of the need for innovative security algorithms with improved payload and imperceptibility.

The organization of the subsequent sections of this chapter is as follows. The preliminary information is given in Section 1.2. In Section 1.3, several techniques for tamper detection and localization have been discussed. The several types of image tampering and attacks that might occur, as well as how digital watermarking helps to reduce them, are also examined. In Section 1.4, the comparison outcomes are displayed. In Section 1.5, the chapter is concluded.

Table 1.1 Existing studies on tamper detection using digital watermarking

Reference	*Methods used*	*Description*
[13]	Arnold transform, DCT operation, SHA-256 hash function	A hash-based fragile watermarking technique is utilized to preserve authenticity and identify altered areas
[12]	Chaotic system–based logistic map	In order to conduct tamper detection and localization in a picture, two watermarking algorithms were developed in this work. The suggested methods employ blindness, and both at the transmitter and receiver ends, the watermark bits are produced using chaotic systems–based logistic maps
[14]	K-means clustering algorithm	The technique is based on utilizing the DCT coefficients to authenticate each 8 × 8 picture block. For each 2 × 2 picture sub-block, they suggested a recovery method that makes use of the K-means clustering algorithm
[15]	SHA-1 hashing algorithm	A block-based watermarking method utilizing the SHA-1 hashing algorithm was presented. Detecting and locating tamper in greyscale photos was the goal of the proposed approach
[16]	Two-level DWT, Schur decomposition, quantum Hilbert image scrambling	This study presents a watermarking method based on DWT and Schur decomposition to detect tampering. The LL sub-band coefficients are taken into consideration while embedding the watermark. The LSB of each block is made 0, and the authenticated block bits (ABB) are generated using the Schur decomposition technique
[17]	Lossy compression approach, CDF9/7 biorthogonal wavelet, SPIHT	This work presents a watermarking system employing lossy compression for forgery detection. The process comprises five steps: employing wavelet transform with biorthogonal CDF 9/7, encoding wavelet coefficients using the SPIHT technique, rearranging watermark bits, encrypting with private keys, generating two tamper detection bits, and embedding them into the cover image (CI)
[18]	Scrambling through the logistic chaotic map, encoding using BCH, QIM mechanism	The proposed method involves extracting transform coefficients from fragmented audio through wavelet and cosine transforms. It utilizes logistic map and BCH coding for watermark image pre-processing, followed by combining the pre-processed image with audio signal coefficients' singular values. Hashes are then generated and inserted into watermarked audio frames, enabling the receiver to recover hashes and identify any audio alterations
[7]	DWT, LSBs	Initially, a watermark is formed from the image itself that is sensitive to content alteration as well as resistant to standard image processing. The created watermark is included in the picture. The image's watermark is then removed in order to locate any tampering, detect it, and restore the image as closely as possible to the original

1.2 BACKGROUND

The security and integrity of data can be ensured via digital encryption (i.e., cryptography), steganography, and digital watermarking. To safeguard data, cryptography uses a variety of codes. The impression of the data itself is changed by encryption techniques. The encrypted data can only be unlocked with a genuine key. Steganography is another way of data security. Secret writing is referred to as steganography. A message can be concealed within another signal, such as audio, video, or picture file, using a technique called steganography. Confidential information dissemination and concealment are two of steganography's main applications. On the other hand, watermarking is a different way to stop data from being abused. The security based on watermarking also has the added benefit of maintaining the original data's shape. Digital watermarking is possible for many forms of data, including pictures, music, and videos. The terminology "digital watermarking" refers to the use of watermarking to protect the integrity of electronic media (such as photos, audio, and video). In addition to other crucial uses like copyright protection, fingerprinting, broadcast monitoring, etc., digital watermarking is used to detect media manipulation and, on rare occasions, to recover lost data from the medium. With a reliable tamper detection system, the altered region in a medium can potentially be localized.

Digital watermarking may be categorized into three categories: spatial, transform, and hybrid. By using decomposition techniques like LU, QR, SVD, Schur, etc., image pixel values are taken into account in the spatial domain. This domain is easy to use, but it is also quite vulnerable to attack. When using several transformation methods in the transform domain, such as the DCT, discrete sine transform (DST), DWT, and contourlet transform, the coefficients of the image are taken into consideration. A number of transformation techniques are used in the hybrid domain, which is an extension of the transform domain.

Based on human perception, watermarks might be visible or not. A conspicuous watermark is a practice of superimposing a mark or data visibly over the CI to produce copyright identification. Typically, a visible watermark is a company logo or phrase that may be seen next to the host image. In contrast, an invisible watermark uses the information enshrined inside the image as a watermark. A corporate logo, a linguistic statement indicating the provenance of the image, a fraction of the CI, or even a total replica of the CI itself might be used as an unnoticeable watermark. There are various invisible watermarking techniques, which are divided into three categories: (i) robust, (ii) semi-fragile, and (iii) fragile.

A robust watermark is designed to withstand several efforts to erase the image's watermark. A robust watermark's principal aim is to verify the authenticity or secure copyright. A robust watermarking strategy may survive a wide range of destructive attacks like compression, cropping, scaling, rotation, and several others. Robust and fragile watermarking methods are combined to create semi-fragile watermarking schemes. Semi-fragile watermarks may identify damaged areas in a photograph and remove the residual watermark from undamaged portions. As a result, even after a subject has been altered, a semi-fragile watermark may still be used to verify it. Ultimately, with even minor changes to the watermark, the fragile watermarking process cannot retrieve the watermark. As a result, they are frequently used for authentication, content integrity, copyright protection, tamper detection, and localization. Digital watermarking may be categorized into three groups based on extraction: (i) semi-blind, (ii) non-blind, and (iii) blind. In blind watermarking, a picture can be watermarked without requiring the original image's metadata, but a semi-blind image needs a fraction of the original image's data. A detailed classification of digital watermarking based on different criteria along with some application areas is shown in Figure 1.1.

The identification of media tampering is one use for digital watermarking. A digital media file can be protected with a watermark, which can be extracted and compared to the original watermark to reveal any alterations or tampering. The file has probably been altered if the retrieved watermark does not match the original. As a result, digital watermarking is crucial for media tamper detection since it offers a means of confirming the validity and integrity of digital media files.

A simplistic tampering assault on a picture could be the insertion of an object or a word. It's possible to alter the image in several ways, including adding a signature or changing the face in an image. All of these are common tampering attacks. Crop attack is another common picture manipulation technique. A section of the image is eliminated in a crop attack. Cropping is divided into three subcategories: (i) Cropping as a full, in which only one chunk is removed from the picture. (ii) Multiple cropping, which incorporates spread distributed cropping, in which the removal of the cropping is distributed throughout the image. (iii) Chunk distribute cropping, in which a few, comparatively large chunks are removed from the image.

Another well-known picture-altering exploit is copy–paste tampering. In copy–paste tampering, a portion of another watermarked picture produced using the identical watermarking algorithm is transferred onto a different image. Further, a copy–move assault is similar to a copy–paste assault. Even the phrase "copy–move forgery" is used to describe this. A copy–move attack entails copying a fragment of the same watermarked image (WI) onto the image itself. Many implementations use images as the watermark; these systems are prone to constant average attacks. The constant average attack begins with an item being added or deleted from the picture. The average of the tampered region is then calculated and combined with the tampered region. A watermarking algorithm should be able to identify such counterfeiting. Different types of tampering attacks, be they intentional or unintentional, are shown in Figure 1.2.

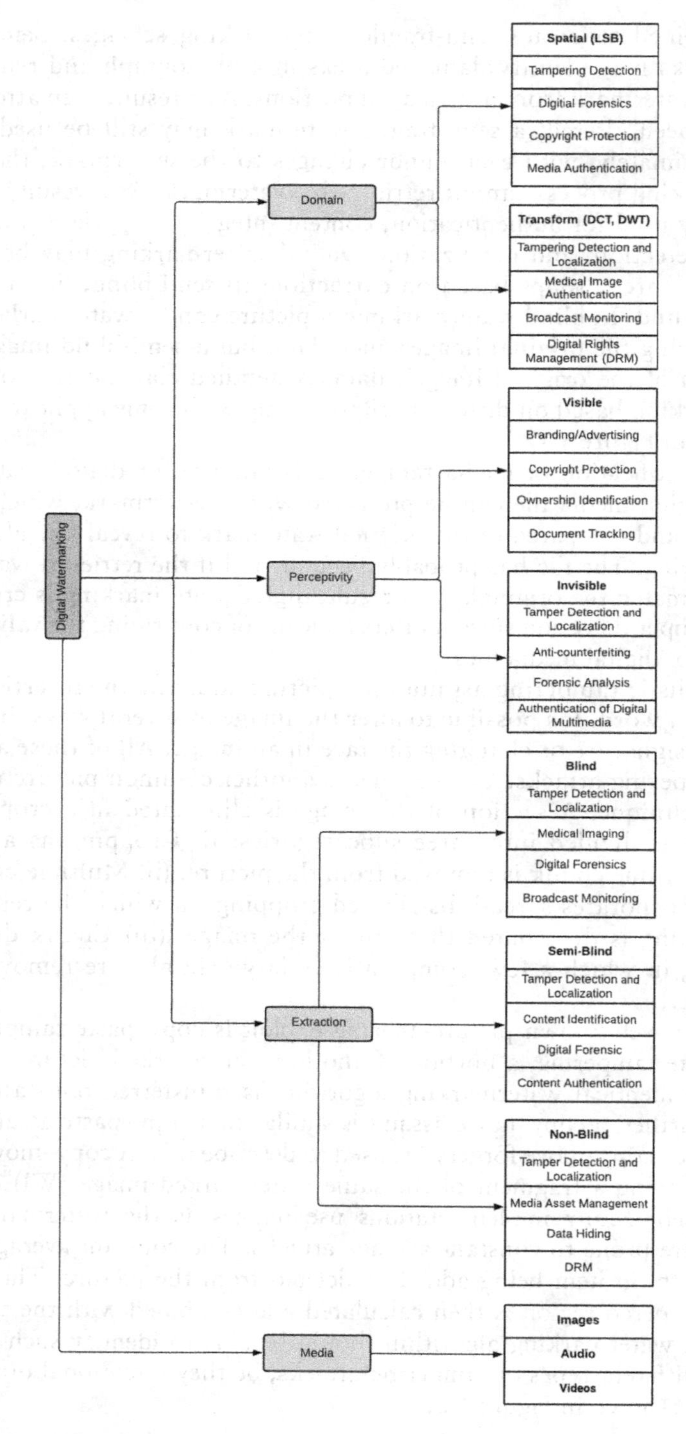

Figure 1.1 Classification of digital watermarking and their applications.

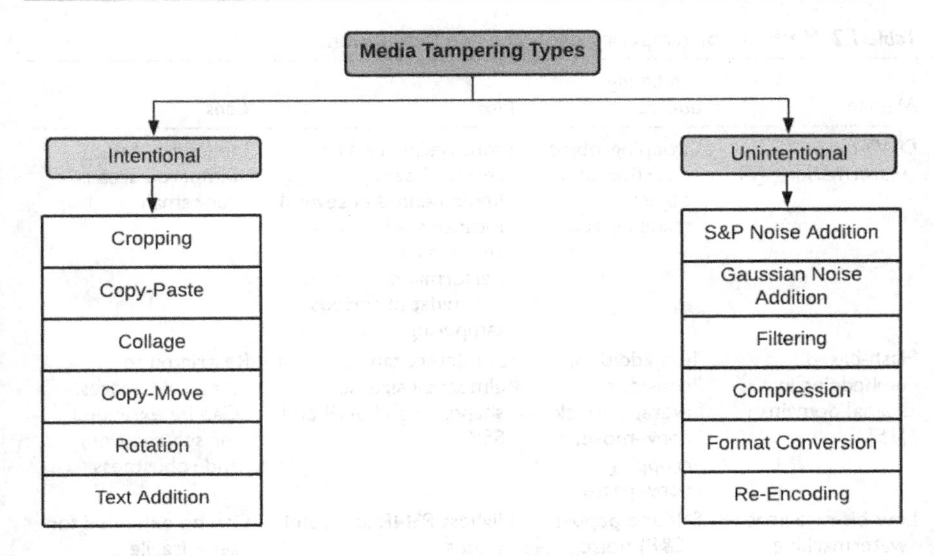

Figure 1.2 Classification of media tampering types.

1.3 METHODS OF MEDIA TAMPER DETECTION AND LOCALIZATION USING DIGITAL WATERMARKING

The various methods used for media temper detection and localization are portrayed in Table 1.2 with their pros and cons.

1.3.1 DWT-based watermarking

A 1D DWT-based watermarking model for picture tamper detection, localization, and restoration has been proposed by Som et al. [7] in their research. Their plan involves three distinct phases. Initially, following preparation for doing so, a watermark that is sensitive to content modification and unaffected by conventional image processing is created from the image source. The created watermark is then included in the picture. Lastly, the watermark is removed from the image, which has undergone much mortification as a result of noise and/or cropping assaults, in order to identify and locate tampering and restore the image as closely as possible to the authentic one. The proposed solution has been put to the test against various cropping assaults. The efficiency of the recommended method is evaluated using the peak signal-to-noise ratio (PSNR) and the structural similarity index (SSIM), which is shown in Table 1.3.

An alternative paradigm is put up in Ref. [16], where two-level DWT-based watermarking for tamper detection of images has been used in the medical field. Sanivarapu [16] used the DWT technique to incorporate the watermark information into meaningful pictures, the quantum Hilbert image scrambling technique to scramble the watermark to yield security, and the Schur

Table 1.2 Methods for tampering detection and localization

Method	*Tampering attacks*	*Pros*	*Cons*
DWT-based watermarking [7]	Cropping, object insertion, and object manipulation	More resilience as a result of being implemented in several locations, which has aided in its performance even in the midst of serious tampering	Ineffective if the tampered area is very small
Hash-based embedding in spatial domain [15]	Text addition, constant average attack, copy–move, cropping, copy–paste	Can detect tampering of almost all sizes and shapes. High PSNR and SSIM	Restricted to greyscale images. Can be extended for self-recovery and robustness
Low-bit-rate image watermarking [17]	Salt and pepper (S&P) noise, cropping, copy–move	Highest PSNR and SSIM values	Can be extended for semi-fragile watermarking coupled with ML/DL algorithm
Hash-based watermarking [13]	Image processing, copy–move, copy–paste	Better PSNR and SSIM values. Computational complexity is good	Complexity, embedding capacity, and perceptual capacity can be enhanced. Can be extended for self-recovery
Logistic map–based blind and fragile watermarking [11]	Rotation, cropping, S&P noise, collage, copy–paste, Gaussian noise	Less execution time, marginal complexity, impressive performance against a variety of attacks	Can be extended for self-recovery, robustness can be improved by using hashing and hamming codes
Reversible and irreversible fragile watermarking using logistic map [12]	Cropping, addition of S&P noise, copy–paste, collage	Outstanding results against attacks, low computational complexity	Self-recovery of tampered bits is not explored, both spatial and transform domains can be augmented to enhance robustness

decomposition technique to construct ABB to detect tampering with medical images. The technique uses a two-level DWT to extract four sub-bands ($C_{LL}, C_{LH}, C_{HL}, C_{HH}$) from a relevant image. While embedding the watermark, the coefficients of LL sub-bands are taken into account. To fend off assaults using image processing, the SB (Sub-bands) is divided into blocks. Each

Table 1.3 Comparison of PSNR and SSIM using various proposed models

Reference	*Methods*	*Image*	*PSNR*	*SSIM*
[7]	DWT-based watermarking	Lena	41.44	0.93
		Peppers	41.39	0.93
		Baboon	41.31	0.98
		Boat	41.32	0.95
[15]	Secure and efficient fragile image watermarking (spatial domain, hash-based embedding)	Lena	51.1246	0.9969
		Barbara	51.1366	0.9977
		Plane	51.1453	0.9963
		Peppers	51.1391	0.9968
		Medical	51.1101	0.9957
[17]	An efficient low-bit-rate image watermarking	Lena	56.1502	0.9989
		Peppers	56.5993	0.9985
		Airplane	56.5207	0.9967
		Baboon	56.212	0.9998
[13]	Hash-based watermarking	Lena	51.14	0.9983
		Sailboat	51.17	0.9987
		Plane	51.19	0.9981
		Baboon	51.23	0.9993
[11]	Logistic map-based blind and fragile watermarking	average of chosen 15 images	51.14	0.9969
[12]	Reversible and Irreversible Fragile watermarking using Logistic map	average of chosen 12 images	54.18	0.9985

block's LSB is set to zero, and then the ABB are generated via Schur decomposition. Quantum Hilbert image scrambling is used to sabotage the intrusion by creating confusion for the invaders. When an important image has been altered, watermarking aids in authenticity and tamper detection. The tamper detection technique ensures imperceptibility, security, and great robustness against image processing assaults and manipulation, according to experimental results. When there are no assaults, the PSNR values are higher than 30 dB and the NCC values are closer to 1. NCC values with attacks greater than 0.95 show the reliability of the proposed approach.

1.3.2 Hash-based embedding in spatial domain

The watermarking algorithm developed by Bhalerao et al. [15] seeks to identify and locate image tampering. Spatial domain, block-based embedding is the approach that is being suggested. It introduces a distinctive key-based embedding that uses SHA-1 hashing. For image tampering detection, the suggested method employs a key-based authentication scheme. The CI is initially broken up into 4×4 non-coincidental blocks. Each block was given a 16-bit block key to identify it. A special block key was used to create the watermark for the block, which aided with tamper detection. An essential requirement for successful extraction is that the keys created for each block should be the same during embedding as well as extraction. As a result, block keys were produced at both ends with the identical seed.

Each block went through the embedding process. The initial host block is a 16-pixel, 4×4 non-overlapping rectangle. The LSB of each pixel was replaced with a zero, and then the SHA-1 hash was produced. The generated hash-key code was stored within the block itself. Each block was embedded, and then the picture was repacked.

Throughout the extraction procedure, the image was once more broken up. Each block key was produced using the same embedding seed. The same key was created for each block throughout embedding because the seed was the same for both embedding and extraction. The LSBs of the block's 16 pixels were all deleted and changed to 0. The retrieved bits were then embedded with hash-key code.

Each block's hash computation was done after the LSBs were set to zero. To obtain the hash-key code for blocks of WI data, the block keys and block hash codes were XORed. When it was determined that the created code and embedded code were identical, it was designated as a non-altered block.

1.3.3 Low-bit-rate image watermarking

For the purpose of ensuring authenticity and spot counterfeiting, Kabir [17] recommends a watermarking system based on a lossy compression technique. In the suggested method, the watermark bit creation and embedding procedures happen in the transform domain and the spatial domain, respectively. The proposed method consists of five steps which are stated in Algorithm 1.

Algorithm 1: Generalize embedding procedure

1. Wavelet transforms using the biorthogonal CDF 9/7.
2. Using the SPIHT (Set Partition in Hierarchical Tree) technique to encode the wavelet coefficient.
3. Private keys are used for encryption and permutation of the watermark bits.
4. Produce two bits for tamper detection.
5. The cover picture has been embedded at this point.

Additionally, error correction coding is employed to make the proposed method more resistant to various assaults. The extraction process of the watermark makes use of watermark reconstruction, tamper detection, and localization. These steps involved in extraction are stated in Algorithm 2.

Algorithm 2: Common extraction procedure

1. The WI and any other suspicious images have been separated into X × X nonoverlapping image chunks.
2. The retrieved pixel location's whole bitstream has been obtained. Those are the encrypted watermarks.
3. To get the watermark bits back, the extracted watermark must be decrypted.
4. The wavelet approximation coefficients are produced by the SPIHT decoding process, which is applied to the watermark.
5. The closest watermark image is discovered using the inverted wavelet transform.
6. The two-tamper detection bit Ga1, Ga2 is computed for tamper detection and localization.
7. The watermarked picture is used to recover the tamper detection bits a1 and a2. The block is identified as a tampered block if Ga1 = a1 and Ga2 = a2. Otherwise, the block is genuine.

1.3.4 Hash-based watermarking

This method proposes a novel strategy for the localization and detection of tampering in digital photographs using hashing and watermarking. Hussan et al. [13] have suggested using a fragile hash-based watermarking method to pinpoint altered areas of digital photographs and maintain their integrity and authenticity. At first, four distinct, non-overlapping sections of the greyscale images are originally separated from one another. Then, using DCT to retrieve the DC coefficients for each 4 × 4 block. The DC coefficients are hashed using the SHA-256 hash algorithm to generate a 16-bit watermark. Arnold transform has been applied to the original CI. and then cut into 4 × 4 portions in order to include the obtained hash. The image's blocks are jumbled, and the 16-bit hash value is included using the Arnold transform. Finally, the WI is generated using the inverse Arnold transform.

A hash is derived from the image itself and used like a fragile watermark for detecting the tamper which means that even a small change to the embedded watermark might have a big impact. The same key that was used for embedding is also utilized to scramble the watermarked picture at the receiving end. The data is recovered from the LSBs of each block of the jumbled picture after the image is divided into 4 × 4 non-coincidental blocks. In the following phase, the picture blocks are subjected to a DCT procedure. By performing an SHA-256 operation on the blocks' DC coefficients, a hash value is produced. The block is designated as genuine or unaltered if the extracted data and the acquired data are identical. The whole procedure is examined for each of the image's building elements. It is discovered that the suggested solution is resistant to various copy–paste, copy–move, and image processing attacks.

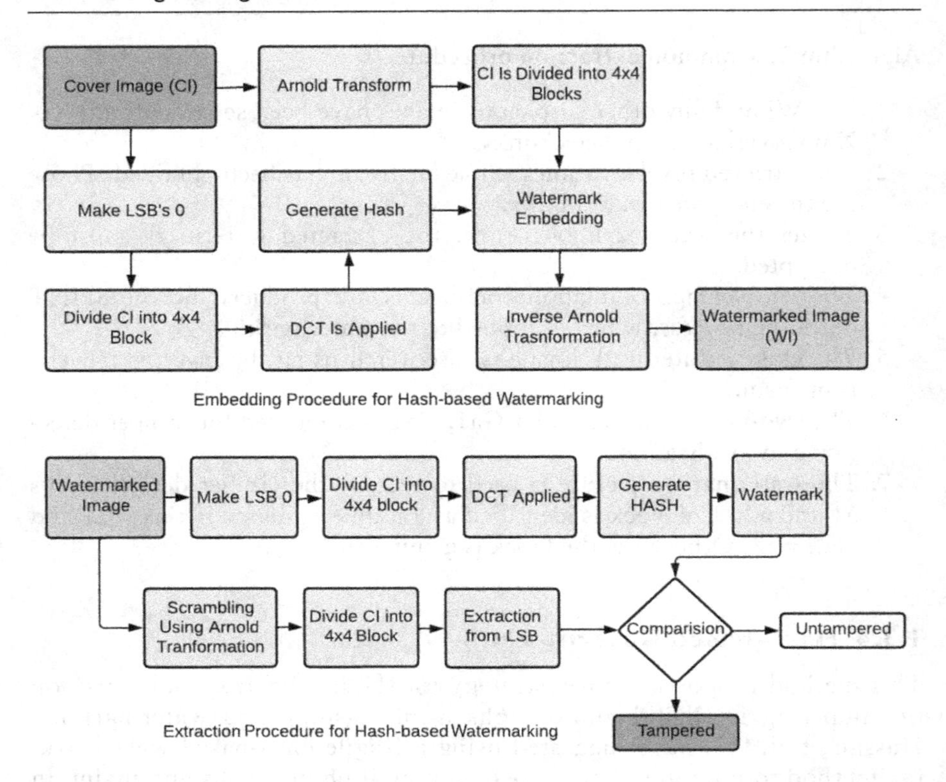

Figure 1.3 Flowchart showing embedding and extraction procedure of hash-based watermarking [14].

The suggested method demonstrates improved imperceptibility with a PSNR of 51.16 on average. The flowchart for the embedding and extraction procedure is shown in Figure 1.3.

1.3.5 Logistic map–based blind and fragile watermarking

The suggested method uses blind watermarking, so the receiver doesn't rely entirely on the WI to get the watermark bits. The suggested approach by Sahu [11] offers good visual quality by embedding the watermark bits in the rightmost bits of each pixel. Regarding the identification of tampered regions and resistance to different picture alteration attempts, the approach can demonstrate excellent performance. The pixels are first converted to an 8-bit binary during the embedding of the watermark, and then the watermark bits along with the initial ISB of each CI's pixel are added as binary. The produced value is next XORed with each pixel's ISB bits. The LSB of the CI pixel receives the result of this XOR operation in the end. The watermarked pixel is made using the suggested way by appending a maximum of 1 to the CI pixel.

On the extraction side, create an equivalent 8-bit binary representation of the watermarked pixels from the ISB and LSB bits of watermarked pixels, and then use the XOR function to join these two bits. Thereafter, the watermark bit is obtained by binary subtraction of the resulting bit from the ISB of the appropriate watermarked pixel.

1.3.6 Reversible and irreversible fragile watermarking using logistic map

The first system that is being suggested is a pixel-by-pixel irreversible fragile watermarking system [12]. One watermarking bit is included in each HI pixel. A block of two consecutive HI pixels is taken into account for the implantation of two watermark bits. The relevant watermark bits are then compared to the 8 bits of each HI pixel. If the eighth bit of each watermark pixel matches the appropriate embedding bit, then the values of the watermark pixels will always remain the same as the HI pixel. Otherwise, if the second watermarked pixel's eighth bit is different from the first, its value is decreased by 1.

Similar to this, if the first embedding bit's 8-bit is different from the first watermark bit and the last HI pixel value is 0, the value of the first watermarked pixel is raised by 1. Concurrently, depending on whether the second pixel's eighth bit matches the embedding bit or not, the watermarked pixel's value will increase by 1. Last but not least, if the first embedding bit's 8-bit differs from the first watermark bit's 8-bit, the first watermarked pixel's value is reduced by 1. Nevertheless, if somehow the second pixel's eighth bit matches the embedding bit, the watermark pixel's value will remain static; or else, it will get raised by 1.

The second technique is reversible [12]. In this case, a watermarking bit has been added to each HI pixel. Thereafter, each pixel is replicated during embedding to produce the mirrored pixels. The mirror image's pixel values match those of the HI pixel exactly. The mirrored pixels now cover one watermark bit on each pixel. The embedding bits and the eight similar bits of the mirrored pixels are contrasted. The watermarked pixels for the mirrored pixels are created using a1 embedding method. The HI pixels may also be extracted using the floor value of the average of each pair of watermarked pixels at the receiving end.

1.4 RESULT AND ANALYSIS

When comparing watermarked photos to original photos, it is crucial to consider the perceived quality and integrity of the result. The ratio of the largest possible mean square difference between any two photos to the mean square difference of any two photos is used to compute an image's PSNR.

Decibels are used to quantify it. The PSNR value must be 30 dB or above, it is generally agreed, for a photo to be considered high quality [19,20]. When two photos are compared, SSIM establishes how similar or different the two images are [21,22]. PSNR should be greater and SSIM values should be near unity for watermarked images. Measurements using PSNR and SSIM are made for watermarked and tampered media [23–25]. Equation (1.1) is a definition of these prohibitions:

$$PSNR = 10 \cdot \log_{10}\left(\frac{255^2}{MSE}\right) \tag{1.1}$$

Here, MSE is given by

$$MSE = \frac{1}{PQ}\sum_{x=1}^{P}\sum_{y=1}^{Q}[H(x,y) - M(x,y)]^2 \tag{1.2}$$

- P and Q denote the rows and columns of the CI, respectively.
- $H(x, y)$ denotes the original image's pixel at (x, y).
- $M(x, y)$ denotes the embedded image's pixel at (x, y).

Further, these three terms – luminance, structure, and contrast – are used to locate the SSIM. SSIM is considered to be

$$SSIM(S,U) = [f(S,U)]^{\alpha} \cdot [g(S,U)]^{\beta} \cdot [h(S,U)]^{\gamma} \tag{1.3}$$

where

$$f(S,U) = \frac{2\mu_s\mu_u + c_1}{\mu_s^2 + \mu_u^2 + c_1} \tag{1.4}$$

where μ_s, μ_u are the mean values of S and U, respectively, and c_1 is a small constant added for numerical stability.

$$g(S,U) = \frac{2\sigma_{s,u} + c_2}{\sigma_s^2 + \sigma_u^2 + c_2} \tag{1.5}$$

where σ_s, and σ_u are the standard deviations of S and U, respectively, and c_2 is a small constant added for numerical stability.

Also,

$$h(S,U) = \frac{\sigma_{su} + c_3}{\sigma_s\sigma_u + c_3} \tag{1.6}$$

where, σ_{su} is the covariance between S and U, and c_3 is a small constant added for numerical stability. Therefore, the SSIM equation is given as follows:

$$SSIM(S,U)=\frac{(2\mu_s\mu_u+C_1)(2\sigma_{su}+C_2)}{(\mu_s^2+\mu_u^2+C_1)(\sigma_s^2+\sigma_u^2+C_2)} \tag{1.7}$$

The comparison of PSNR and SSIM using various suggested models for picture watermarking is shown in Table 1.3. The table lists the various strategies employed in the suggested models and offers a quantitative evaluation of their effectiveness using the two well-liked measures: PSNR and SSIM. The findings demonstrate that some of the suggested models, including the effective low bit rate watermarking model and the hash-based watermarking model, have attained greater PSNR and SSIM values than others. It is crucial to remember that the effectiveness of these models can change based on the particular image being used. Overall, the table is a useful resource for academics and professionals looking to investigate the efficacy of various picture watermarking methods.

1.5 CONCLUSION

In this era of digital communication, the security of multimedia transmissions is now of extreme significance. Thus, in this chapter, a detailed study of several methods to prevent the tampering of digital media and to localize the tampered area has been presented. The tampering of digital media has become very easy with the availability of image processing tools, and hence digital watermarking is found as a solution to detect altered media. In this chapter, several tampering types or attacks are presented, and to prevent those attacks, various methods have been discussed. The ability of various methods is compared on the basis of PSNR and SSIM values. Through this study, several key findings have been identified by us, such as the highest PSNR and SSIM values of 56.599 and 0.999, respectively, are achieved by low-bit-rate image watermarking. Also, the hash-based watermarking provides good PSNR and SSIM values of 51.19 and 0.999, respectively. These methods have been experimented against several intentional attacks such as cropping, copy–paste, copy–move, etc. Some methods, like low-bit-rate watermarking and logistic map-based watermarking, are tested against unintentional attacks such as S&P noise, Gaussian noise, and more. It is concluded that the suitable method for tamper detection and localization is low-bit-rate image watermarking followed by hash-based watermarking and others.

However, in the future, these methods can be extended for even better results, higher PSNR and SSIM values, and less computational complexity. Also, self-recovery of the tampered region is an aspect that is yet untouched. Overall, this chapter highlights the need for further research and development of new and efficient methods which can detect the tampering more accurately, localize them, and provide self-recovery for the tampered region. Researchers, professionals, and decision-makers who are interested in expanding knowledge in this field may find this chapter to be a useful resource.

REFERENCES

1. Archana Tiwari, Manisha Sharma, and Raunak Kumar Tamrakar. Watermarking based image authentication and tamper detection algorithm using vector quantization approach". In: *AEU: International Journal of Electronics and Communications*, 78 (2017), pp. 114–123.
2. Faranak Tohidi and Manoranjan Paul. "A new image watermarking scheme for efficient tamper detection, localization and recovery". In: *2019 IEEE International Conference on Multimedia & Expo Workshops (ICMEW)*. IEEE. 2019, pp. 19–24.
3. Sripradha, R. and Deepa, K. "A new fragile image-in-audio watermarking scheme for tamper detection". In: *2020 3rd International Conference on Intelligent Sustainable Systems (ICISS)*. IEEE. 2020, pp. 767–773.
4. Sami Bourouis et al. "Recent advances in digital multimedia tampering detection for forensics analysis". In: *Symmetry* 12.11 (2020), p. 1811.
5. Chennamma, H. R. and Madhushree, B. "A comprehensive survey on image authentication for tamper detection with localization". In: *Multimedia Tools and Applications* 82.2 (2023), pp. 1873–1904.
6. Shiv Prasad and Arup Kumar Pal. "A tamper detection suitable fragile watermarking scheme based on novel payload embedding strategy". In: *Multimedia Tools and Applications* 79.3–4 (2020), pp. 1673–1705.
7. Sukalyan Som et al. "A DWT-based digital watermarking scheme for image tamper detection, localization, and restoration". In: *Applied Computation and Security Systems: Volume Two* (2015), Springer. pp. 17–37.
8. Solihah Gull et al. "An efficient watermarking technique for tamper detection and localization of medical images". In: *Journal of ambient intelligence and humanized computing* 11 (2020), pp. 1799–1808.
9. Mahboubeh Nazari, Amir Sharif, and Majid Mollaeefar. "An improved method for digital image fragile watermarking based on chaotic maps". In: *Multimedia Tools and Applications* 76 (2017), pp. 16107–16123.
10. Saswati Trivedy and Arup Kumar Pal. "A logistic map-based fragile watermarking scheme of digital images with tamper detection". In: *Iranian Journal of Science and Technology, Transactions of Electrical Engineering* 41 (2017), pp. 103–113.
11. Aditya Kumar Sahu. "A logistic map based blind and fragile watermarking for tamper detection and localization in images". In: *Journal of Ambient Intelligence and Humanized Computing* 13.8 (2022), pp. 3869–3881.
12. Aditya Kumar Sahu et al. "Logistic-map based fragile image watermarking scheme for tamper detection and localization". In: *Multimedia Tools and Applications* 82.16 (2022), pp. 1–32.
13. Muzamil Hussan et al. "Hash-based image watermarking technique for tamper detection and localization". In: *Health and Technology* 12.2 (2022), pp. 385–400.
14. Assem Abdelhakim, Hassan I Saleh, and Mai Abdelhakim. "Fragile watermarking for image tamper detection and localization with effective recovery capability using K-means clustering". In: *Multimedia Tools and Applications* 78 (2019), pp. 32523–32563.
15. Siddharth Bhalerao, Irshad Ahmad Ansari, and Anil Kumar. "A secure image watermarking for tamper detection and localization". In: *Journal of Ambient Intelligence and Humanized Computing* 12.1 (2021), pp. 1057–1068.

16. Prasanth Vaidya Sanivarapu. "Adaptive tamper detection watermarking scheme for medical images in transform domain". In: *Multimedia Tools and Applications* 81.8 (2022), pp. 11605–11619.
17. Md Ahasan Kabir. "An efficient low bit rate image watermarking and tamper detection for image authentication". In: *SN Applied Sciences* 3.4 (2021), p. 400.
18. Venkata Lalitha Narla et al. "BCH encoded robust and blind audio watermarking with tamper detection using hash". In: *Multimedia Tools and Applications* 80.21–23 (2021), pp. 32925–32945.
19. Deb, S., Das, A., & Kar, N. "An applied image cryptosystem on Moore's automaton operating on δ (qk)/𝔽 2". In: *ACM Transactions on Multimedia Computing, Communications and Applications* 20.2 (2023), pp. 1–20.
20. Asmitha, P., Rupa, C., Nikitha, S., Hemalatha, J., and Sahu, A. K. "Improved multiview biometric object detection for anti spoofing frauds". In: *Multimedia Tools and Applications* (2024), pp. 1–17. https://doi.org/10.1007/s11042-024-18458-8
21. Sahu, A. K., Umachandran, K., Biradar, V. D., Comfort, O., Sri Vigna Hema, V., Odimegwu, F., and Saifullah, M. A. "A study on content tampering in multimedia watermarking". In: *SN Computer Science* 4.3 (2023), p. 222.
22. Kamil Khudhair, S., Sahu, M., Raghunandan, K. R., and Sahu, A. K. "Secure reversible data hiding using block-wise histogram shifting". *Electronics* 12.5 (2023), p. 1222.
23. Raghunandan, K. R., Dodmane, R., Bhavya, K., Rao, N. K., and Sahu, A. K. "Chaotic-map based encryption for 3D point and 3D mesh fog data in edge computing". *IEEE Access* 11 (2022), pp. 3545–3554.
24. Roy, K. S., Deb, S., and Kalita, H. K. "A novel hybrid authentication protocol utilizing lattice-based cryptography for IoT devices in fog networks". *Digital Communications and Networks* 10.4 (2022), pp. 989–1000. https://doi.org/10.1016/j.dcan.2022.12.003
25. Deb, S., Pal, S., and Bhuyan, B. "NMRMG: nonlinear multiple-recursive matrix generator design approaches and its randomness analysis". *Wireless Personal Communications* 125.1 (2022), pp. 577–597.

Chapter 2

Implementation and analysis of digital watermarking techniques for multimedia authentication

Sayan Das, Priyanka Biswas, Nirmalya Kar, and Aditya Kumar Sahu

2.1 INTRODUCTION

Since the introduction of the internet, the amount of multimedia content being used practically has been increasing exponentially. A lot of the multimedia content that we consume daily is from authentic sources, that is, sources that can be trusted to provide us genuine content and not artificial intelligence (AI)–generated fictional pictures which are very much difficult to differentiate from original these days at plain sight. However, it is getting extremely difficult to point out the authenticity of a particular content with the increase in open-source AI tools.

Techniques for source origin authentication and multimedia integrity verification are referred to as "multimedia authentications." Digital watermarking [1] or digital signatures [2] are used to apply the procedures. A digital signature is an encrypted message digest that is non-repudiable and retrieved from the content. It is typically kept in a separate file that may be connected to the data to demonstrate its originality and integrity. Digital watermarking, on the other hand, adopts the strategy of incorporating the watermark into the multimedia data such that it is present in the protected multimedia data.

Digital watermarking can be applied to various types of multimedia data, including images, audio, and video. However, different types of multimedia data require different watermarking techniques. For example, audio watermarking requires techniques that are sensitive to changes in the audio signal, while video watermarking must take into account the complex nature of video data, including motion, lighting, and color. Digital watermarking's primary goal is to safeguard the veracity and consistency of multimedia data. The watermarking process involves the insertion of the watermark into the multimedia data, which can later be extracted to verify the authenticity of the data.

Authentication [3] using digital watermarking can be achieved in several ways. One common method is to embed a watermark that is resistant to different kinds of manipulations, such as signal processing, compression, and format conversion. This type of watermark is called a robust watermark and is usually used for copyright protection and content authentication.

 DOI: 10.1201/9781032663647-2

2.2 BACKGROUND

Before getting into the different watermarking methods utilized in authentication, the following sections offer an introduction to digital watermarking and authentication.

2.2.1 Digital watermarking

The process of concealing or surreptitiously encoding digital information in a digital signal is known as digital watermarking. The digital watermark is the name of the embedded signal. The host or carrier signal is the transmission that contains the watermark. Any sort of signal, including music, video, images, and text, can be the carrier signal.

There are three main steps required to create a working digital watermarking system, namely, watermark generation, embedding, and extraction. Figure 2.1 illustrates the procedure of digital watermarking.

- *Watermark Generation:* This is the step in the digital watermarking process where unique properties of the host signal, be it image, audio, text, or video, are subjected to a series of pseudorandom permutations so as to obtain a watermarked signal that is entirely unique to that particular host. Every image, audio, text, or video when passed through the same watermark generator will generate different watermarks which are unique to that particular image, or host objects of other types.
- *Watermark Embedding:* The watermarked signal generated in the previous step is now embedded carefully in the host media in such a manner that the properties of the host media do not change. Watermarks in practice, for authentication purposes, are mostly invisible. The embedding technique should also ensure that the watermarked signal does

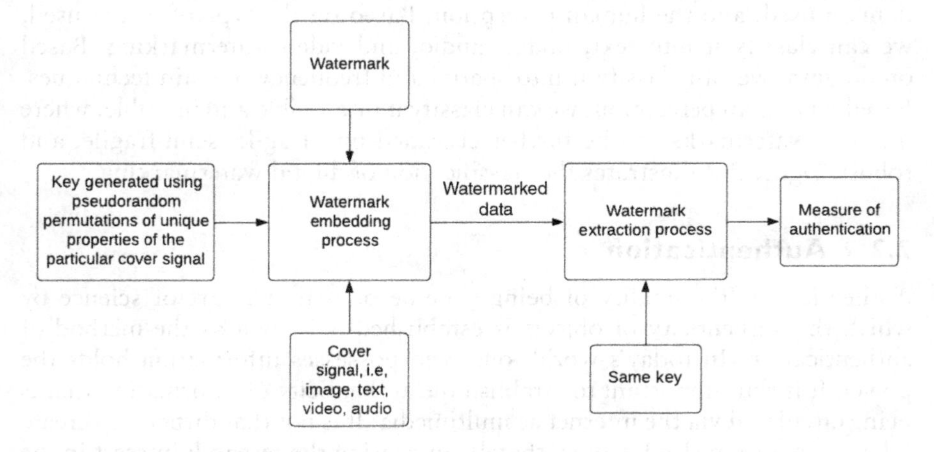

Figure 2.1 Digital watermarking process.

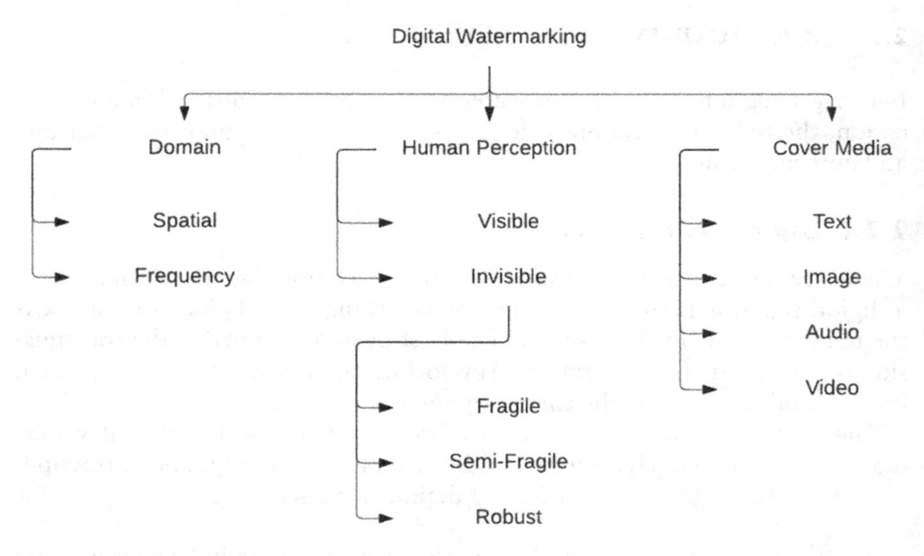

Figure 2.2 Classification of digital watermarking.

not change when subjected to noise, i.e., distortions in the traveling medium.

- *Watermark Extraction:* The opposite of watermark embedding is watermark extraction. The watermark on the cover image can be found and removed using the same secret information (key). The carrier signal can be validated and tampering can be discovered once the watermark has been fully removed and recovered.

Digital watermarking can be classified based on the type of media used, the domain used, and the human perception. Based on the type of media used, we can classify it into text, image, audio, and video watermarking. Based on domain, we can classify it into spatial and frequency domain techniques. Based on human perception, we can classify it into visible and invisible, where invisible watermarks can be further classified into fragile, semi-fragile, and robust. Figure 2.2 illustrates the classification of digital watermarking.

2.2.2 Authentication

Authenticity is the quality of being genuine or real. The art or science by which the authenticity of objects is established is known as the method of authentication. In today's world, one who possesses information holds the power. It is thus important to establish the authenticity of information that is being circulated via the internet as multimedia. It is not that difficult to create fake content to mislead people, thereby increasing the research interest in the field of authentication techniques.

Authentication of multimedia content can be done in two major ways—one of them being digital signature certificate and the other one being digital watermarking techniques. Digital watermarking or digital signatures are used to apply the procedures. A digital signature is an encrypted message digest that is non-repudiable and retrieved from the content. It is typically kept in a separate file that may be connected to the data to demonstrate its originality and integrity. Digital watermarking, on the other hand, adopts the strategy of incorporating the watermark into the multimedia data such that it is present in the protected multimedia data.

Using traditional signatures, even an alteration of a single bit in multimedia data can cause the authentication to be lost. There are two key benefits of digital watermarking for authentication. First, to get around the cryptographic signature attached, the watermark is embedded as a natural component of the host image. Second, since the watermark is hidden within the data, it goes through the same metamorphosis as the data. By looking at the altered watermark, these modifications can be undone.

In this chapter, we are going to focus on authentication techniques based on digital watermarking. The following section will present thorough research on the existing digital watermarking techniques for multimedia authentication.

2.3 DIGITAL WATERMARKING IN AUTHENTICATION

In this section, a thorough description of the existing watermarking techniques is presented on how different cover media, i.e., multimedia content, can be authenticated using digital watermarking techniques.

2.3.1 Image authentication

Establishing the authenticity of images is becoming increasingly difficult these days. Several kinds of image tampering attacks are being performed these days to destroy the ownership information embedded in the image. The internet is also full of new AI-powered tools that greatly help forgers in this regard. Attacks include cropping the image, rotating the image, compressing the image, and adding noise to the image. Table 2.1 provides a systematic description of the mentioned methods. An analysis of the performance of the techniques and how they hold up against some attacks are presented in Section 2.4.

2.3.2 Audio authentication

Audio authentication techniques using digital watermarking are used to verify the integrity and authenticity of audio data. Some of the renowned techniques for this purpose include watermarking in the time domain, frequency

Table 2.1 Image authentication techniques

Reference	*Methods*	*Dataset*	*Measures*	*Description*
[4]	Semi-fragile watermarking technique using vector quantization	Lena, Baboon, Pepper, Cameraman, and F-16	PSNR, NHS	Similarity index between the embedded and the extracted semi-fragile watermark is checked in the proposed authentication mechanism. There is a chance that the received image could be the target of a security attack if the value of NHS = 0.99
[5]	Lifting weight transform followed by discrete cosine transform (DCT)	Baboon, Barbara, Cameraman, Santiago, F-16, Elaine, Pepper, and Lena	PSNR, SSIM	This paper proposes an efficient semi-fragile watermarking method based on LWT and FNN for picture authentication. In this study, the correlation approach is used to embed the random watermark bits into the LWT high-frequency band
[6]	Camouflaged unseen-visible watermarking (CUVW)	UCID Dataset [7]: 1338 uncompressed TIFF color images and RAISE-1k Dataset [8]	PSNR, SSIM, NCD	In order to boost robustness without lowering the quality of the color image, approach embeds the watermark signal utilizing the luminance data from the YCbCr color model and the texture attributes generated by a texture classification method based on DCT and a just noticeable distortion (JND) criterion
[9]	Discrete wavelet transform (DWT)	Airplane, Lena, Peppers, and Barbara	PSNR	To reliably certify manipulated regions in watermarked photos, this research developed a new, reversible watermarking approach. Watermark embedding and extraction are carried out in the frequency domain. The identification code is generated randomly and placed in the second DWT low-frequency subbands of each image block in the proposed approach
[10]	Singular value decomposition (SVD)-based fragile watermarking	12 grayscale medical images	FPR, FNR, PSNR, TDR, NCC	In this proposed scheme, a host image is divided into 4 × 4 blocks, and then SVD is applied by inserting blockwise SVD traces into the image's pixels' least significant bits to determine the original image's transformation. To resist the vector quantization attack, two authentication bits—block authentication and self-recovery bits—are utilized

(Continued)

Table 2.1 Image authentication techniques *(Continued)*

Reference	*Methods*	*Dataset*	*Measures*	*Description*
[11]	Wavelet packets decomposition-based semi-fragile watermarking (generalized DWT)	Wide variety of color images including but not limited to "Lena," "Plane," "Mandrill" "Peppers," etc.	CPSNR, BER, SSIM, MOS	With the ability to precisely detect forgery and recover tampered sections, a unique WPD-based semi-fragile system has been proposed to be applied for copyright protection against various forms of attacks
[12]	Authentication using LSB substitution	"Lena," "SailBoat," "Airplane," and "Goodhill"	PSNR, SSIM	Two authentication bits, a recovery bit to be embedded in the pixel, and two pixel position bits are created for each pixel using the pixel's 5 MSBs
[13]	Two-level wavelet transform on subblocks	Image material including eight gray images: "Sailboat, Lena, Peppers, Baboon, F-16, Boat, Cameraman, Man" and 120 images downloaded from the USC-SIPI Image Database [14] and the ILSVRC2016 dataset [15]	PSNR, SSIM	Suggested method uses wavelet transform in two levels on each of the image's no-overlapping, 16×16 subblocks. Then, according to the key (k1), $n + 2$ data at various points inside the LL2 domain of each subblock are randomly chosen. These data are then utilized for feature extraction and watermark embedding, respectively
[16]	DWT followed by SVD	"Lena, Pepper, Baboon, Boats, Barbara, Avion". Besides, one fragile watermark and two robust binary watermark images	PSNR, SSIM, BER, NC	Dual watermarking scheme proposed: one invisible robust watermark using EWT based on Arnold encryption, and the other fragile watermark for authentication using DWT followed by pixel division into 4×4 and then performing SVD in each block

domain, spread spectrum, and using quantum index modulation (QIM). These techniques can be used for the authentication of audio data in various applications, including audio watermarking, music copyright protection, audio forensic analysis, and audio authentication in security systems. Table 2.2 provides a summary of the techniques used for audio authentication.

Table 2.2 Audio authentication techniques

Reference	*Methods used*	*Dataset*	*Measures*	*Description*
[17]	Audio compression using DCT	100 WAVE format speech signals recorded by AudioRecorder, SonyPCM-D100	SDG, ODG, BER	An effective compression method using DCT for digital speech signal has been proposed. It involves scrambling after sampling each frame and mapping the frame number to embed into the first two segments and embedding the compressed signal in the last two segments for tamper recovery and authentication
[18]	High-order difference statistic	50 wave audio files (bipolar, 16-bit, and sample rate –44.1 kHz)	SNR, BER	A robust, reversible method that constructs a histogram by calculating high-order difference statistic of each frame is proposed. Further, an integer transform on the histogram data is explained which modifies the histogram to hide data bits
[19]	Indirect synchronization by even frame division	A 16-bit quantized clip of pop music from the 1970s, with sample rate – 44.1 kHz.	SNR, BER	Proposes a method that increases the robustness of the watermark by using an indirect synchronization method that uniformly divides the frames and embeds the watermark using the global rather than local features of the audio frame
[20]	Dual-tree complex wavelet transform and Mel-frequency Cepstral coefficients feature detection	Four audio clips: "Danube.wav," "March.wav," "Piano.wav," and "My heart will go on.wav" (16 bits per sample, sampling rate –44.1 kHz and length 15 s)	SNR	In this study, the MFCC features are used to extract the most discriminative features from the audio signal, while the DT–CWT is used to embed the watermark in a robust and imperceptible way

(Continued)

Table 2.2 Audio authentication techniques *(Continued)*

Reference	*Methods used*	*Dataset*	*Measures*	*Description*
[21]	Spread spectrum in discrete Fourier transform	16-bit audio samples of 60 s with a sampling rate of 8 kHz	SNR	In this paper, the cyclic pattern for watermarks has been used, to detect forgeries without consideration of synchronization
[22]	Orthogonal variable spreading factor and QIM	Eight covert messages and five voice files	SNR	A spread text is inserted into the audio recording's wavelet coefficients as part of the proposed fragile watermarking approach for audio authenticity

2.3.3 Video authentication

Video authentication techniques using digital watermarking have gained much attention due to the widespread use of video content over the internet. The most common approach used in video authentication is the DWT [23–25], which decomposes the video frames into frequency subbands to embed the watermark. DWT-based watermarking techniques provide resistance to basic attacks like compression, filtering, and scaling. Other techniques, such as DCT [26], QIM [27], and region-based watermarking, have also been proposed to improve the robustness and security of video authentication. A summary of video authentication techniques is given in Table 2.3.

Table 2.3 Video authentication techniques

Reference	*Methods used*	*Dataset*	*Measures*	*Description*
[25]	DWT domain	Suzie Video (256 × 256 and 150 frames). 32 × 32 QR Code has been taken as the watermark	PSNR, SF	This research presents an invisible video watermarking system using PCA in the DWT domain. The video frames are translated into YUV color space in order to embed the brightness component. Each video frame contains a binary watermark bit embedded in the principal component scores of the LH band. High-similarity factor allows for the detection of the hidden watermark

(Continued)

Table 2.3 Video authentication techniques *(Continued)*

Reference	*Methods used*	*Dataset*	*Measures*	*Description*
[27]	QIM-based semi-fragile watermarking	A data set consisting of four video sequences: Indian cooking, Primitive cooking, LutGaya, and RealBarca	PSNR, BER	For H.264 video authentication, a semi-fragile watermarking approach is given in this study. At the H.264 quantization stage, the proposed approach inserts a semi-fragile watermark into each key frame's integer DCT coefficient. With about 95% accuracy, the system localizes modified video chunks of size 4 × 4
[28]	LSB steganography algorithm to embed blind watermark in key frames	3 MP4 videos of size 3.92 MB, 10 MB, and 17.6 MB, respectively	PSNR, MSE	The system described in this work combines hybrid techniques to encrypt watermarks (message and image) with multiple cipher algorithms (RSA for message watermark and AES for picture watermark), conceal them using LSB steganography in key frames, and protect videos from copyright infringement and establish ownership information
[29]	DWT-based blind watermarking	Video sample consisting of resolution 360 × 240 resized to a resolution of 512 × 256 with frame numbers from 10 to 60, with a total of 274 frames	NC	Proposed a DWT scheme according to which a watermark is broken into many components and placed in the appropriate frames of various situations in the original video. This guarantees that the suggested technique will be resilient to attacks from frame dropping, averaging, swapping, and lossy compression
[26]	DCT domain	One "dynamic" and one "static" grayscale video (256 × 256 and 25 fps)	PSNR	Presents a semi-fragile watermarking system for video authentication that embeds watermarks into video streams in real-time using DCT domain-based technique

(Continued)

Table 2.3 Video authentication techniques *(Continued)*

Reference	*Methods used*	*Dataset*	*Measures*	*Description*
[30]	Watermark bit embedding approach	HEVC encoded videos of HD resolution (1080p and 2160p).	PSNR, BIR	The proposed technique in this work attaches a readable, imperceptible watermark into specified frames of a video clip that is encoded using HEVC, including a blind extraction process
[31]	Fragile watermarking in spatial domain using Arnold Cat Map	Cartoon film of resolution 480 × 720 and 450 frames	–	Proposes a spatial domain fragile watermarking that uses Arnold Cat Map to generate a random image and enciphers the watermark before attaching it to the pixel values of video frames
[24]	DWT- and SVD-based semi-fragile watermarking	Surveillance samples and "camera2. avi, test.avi, akiyo.avi, video1.avi, coastguard. avi, and news.avi"	PSNR, BER	Suggests a semi-fragile watermarking that uses SURF and MSER detectors to generate content-based authentication signatures. After that, an ROI–SVD–DWT-based embedding approach is used to hide the watermark bits in the host video frames. The suggested method guarantees a high level of imperceptibility
[23]	SVD- and DWT-based semi-fragile watermarking	Video samples such as test, camera2, video1, foreman, mobile, table, template	PSNR, BER, NC	In order to authenticate video content in the transform domain (DWT and SVD), a blind semi-fragile watermarking approach was put out in this study. The system begins by generating watermarks using a QR code technique and features that are taken from ROI. The watermark for authentication is then incorporated into the DWT's singular value matrix coefficients with respect to the mid-frequency subbands after being encrypted by the Arnold transform

2.4 RESULTS AND ANALYSIS

2.4.1 Performance metrics

PSNR mostly used as a performance metric for analyzing how the pixel values of images or individual frames of videos are changed after embedding the watermark, the peak signal-to-noise ratio (PSNR) is expressed in Equation (2.1) where MSE is defined as the mean square error value before and after watermark embedding [32].

$$PSNR = 10\log_{10}\frac{255^2}{MSE} \tag{2.1}$$

Equation (2.2) provides the formula to calculate mean square error, where N denotes the number of pixels in the original image, C' is the value of an individual pixel's intensity in the case of an image after watermark embedding, and similarly, Ci is the intensity value for the image before embedding the watermark.

$$MSE = \sum_{i=1}^{N}\frac{(C_i - C_i')^2}{N} \tag{2.2}$$

SSIM is a performance metric used to calculate the similarity factor between multimedia content before and after the watermark embedding. Structural similarity index measure (SSIM) is calculated using Equation (2.3) where $c1 = (k1,L)2$ and $c2 = (k2,L)2$. μx and μy are the mean intensity of I and S, respectively. $\sigma 2$ and σy are the variances of x and y, respectively. σ_{xy} is the covariance of x and y. c_1 and c_2 are the two constant parameters.

$$SSIM(x,y) = \frac{(2\mu_x\mu_y + c_1)(2\sigma_{xy} + c_2)}{(\mu_x^2 + \mu_y^2 + c_1)(\sigma_x^2 + \sigma_y^2 + c_2)} \tag{2.3}$$

Normalized cross-correlation (NC) is a performance metric that is used in template matching. In watermarking scenarios, this is used to measure the resemblance between embedded and recovered watermarks. The formula for calculating NC is given in Equation (2.4) where m, n are coordinates and I and I' are embedded and recovered watermarks.

$$NC = \frac{\sum_{m,n} I_{m,n} I'_{m,n}}{\sqrt{\sum_{m,n} I_{m,n}^2}\sqrt{\sum_{m,n} I'^2_{m,n}}} \tag{2.4}$$

Bit error rate (BER) represents the ratio of erroneous bits to the total transferred bits. Equation (2.5) provides the formula for BER:

$$BER = \frac{Number\ of\ Bit\ Errors}{Total\ Number\ of\ Bits\ Transmitted} \tag{2.5}$$

2.4.2 Analysis of performance metrics using various proposed models

Based on Table 2.4, we can observe that different image authentication techniques perform differently on different datasets. For example, the technique proposed in Ref. [9] achieves a high PSNR value of around 83 for Lena, Peppers, and Airplane datasets, while the technique proposed in Ref. [6] achieves a high SSIM value of 0.9984 on the average of UCID dataset [7].

We can also observe that some techniques achieve very high similarity factors, indicating a high level of similarity between the original and authenticated images. For example, the techniques proposed in Refs. [11] and [12] achieve SSIM values close to 1 for some datasets, indicating a very high level of similarity between the original and authenticated images.

Table 2.4 Performance analysis of image authentication techniques

Reference	*Method*	*Dataset*	*PSNR*	*Similarity factor*
[11]	Wavelet packet decomposition semi-fragile	Lena Plane Peppers Mandrill	45.68 46.06 46.13 45.46	SSIM: 0.9994 SSIM: 0.9996 SSIM: 0.9996 SSIM: 0.9993
[12]	LSB substitution	Lena Sailboat Airplane Goodhill	38.3545 38.3829 38.3545 38.3418	SSIM: 0.926 SSIM: 0.938 SSIM: 0.9382 SSIM: 0.9521
[13]	Wavelet transformation	Lena Sailboat Baboon Peppers Man	40.71 39.03 42.46 39.8 39.41	SSIM: 0.9861 SSIM: 0.9885 SSIM: 0.996 SSIM: 0.9837 SSIM: 0.9862
[16]	DWT followed by SVD	Lena Peppers Baboon Avion Boats Barbara	53.1365 51.9812 49.1216 50.4827 57.0541 53.7674	NC: 1 NC: 0.9443 NC: 1 – – –
[4]	Vector quantization—semi-fragile	Average of several images	42	NHS: 1.0
[5]	LWT followed by DCT	Lena Peppers Baboon Barbara Santiago Cameraman	43.4694 43.281 41.0983 43.8356 40.9769 46.7033	SSIM: 0.9988 SSIM: 0.9987 SSIM: 0.9968 SSIM: 0.9962 SSIM: 0.9961 SSIM: 0.9869
[6]	CUVW method	Average of UCID dataset	62.2517	SSIM: 0.9984
[9]	DWT	Lena Peppers Airplane	83.54 83.52 83.58	– – –

Table 2.5 Performance of discussed techniques with respect to certain attacks

Methods	*Attacks performed*	*Performance*
[11] Wavelet packet decomposition semi-fragile	JPEG and JPEG2000 compression, Gaussian noise, salt and pepper noise, cropping, and forgery attack	Method provides high imperceptibility and robustness against compression and noise addition. It also successfully recovers unauthentic regions and provides high authentication performance against cropping and forgery attacks
[12] LSB, substitution	Cropping, salt and pepper noise, pixel-based attacks	Works extremely well against pixel-based content shifting attacks, also performs fairly against cropping and noise addition attacks and provides good estimation of the original image
[13] Wavelet transformation	Targeted, collage, and forge attacks	Highly robust, resistance against these attacks is very good as seen by the results of this algorithm
[16] DWT followed by SVD	JPEG compression, gaussian noise, salt and pepper noise, median filtering, cropping, and forgery attack.	Provides great robustness against attacks of moderate magnitude, but is still limited against high JPEG compression and a high degree of salt and pepper noise addition.
[4] Vector quantization—Semi semi-fragile	Cropping, cutting, and object addition (text) attacks	With a very low false positive and false negative rate, this technique efficiently detects attacks
[5] LWT followed by DCT	Noise addition, median filtering, luminance, geometric operations, and compression attacks.	Robustness is admissible in common attacks such as noise addition and luminance but in JPEG compression and median filtering attacks, robustness is low
[6] CUVW method	Geometric, distortions, artistic filtering, signal compressing, and noise addition attacks	Retrieves watermark pattern for all tested scenarios and shows admissible robustness against these attacks
[9] DWT	Content tampering and collage attacks.	The proposed scheme has good resistance against these attacks as seen from the experimental results

Table 2.5 summarizes the performance of different digital watermarking techniques against various types of attacks. The techniques are evaluated based on their imperceptibility, robustness, and ability to detect tampering attacks. Overall, the table provides a comparison of the effectiveness of different digital watermarking techniques against various types of attacks.

Table 2.6 presents an analysis of various audio authentication techniques against several attacks. The attacks against which these methods were tested include deletion attack, substitution attack, MP3 compression, filtering, echo

Table 2.6 Analysis of audio authentication techniques using several attacks

Methods	*Attacks*	*Measures*	*Performance*
Audio compression using DCT [17]	Deletion attack	ODG = –0.673, SDG = –0.9237	This method provides good security against normal attacks and is able to recover the lost/tampered bits using compressed signals
	Substitution attack	ODG = –0.692, SDG = –0.8504	
	Insertion attack	ODG = –0.795, SDG = –0.7937	
DTCWT and MFCCF[20]	MP3 compression, filtering, echo addition, geometric distortions	SNR = 26.70, SDG = 0 (Danube) SNR = 26.91, SDG = –0.01 (heart) SNR = 32.22, SDG = 0 (March) SNR = 32.98, SDG = 0 (Piano)	Robust to mentioned attacks up to a certain extent, for example, withstands up to 40% echo addition, filtering till 4 kHz, volume change, and so on
Indirect synchronization [16]	No attack [2]	For quantization length (QL) 0.005, average SNR = 40.124; and for QL = 0.01, average SNR = 34.48	The proposed technique offers very good resistance against noise addition and filtering attacks, but in case of resampling attacks when the number of watermarked bits are increased, the model provides BER greater than 0.20 which is inadmissible
	Gaussian noise	BER = 0 for QL = 0.005 and 0.01	
	Resampling	BER = 0.0712 (avg) for QL = 0.005 and BER = 0.0112 (avg) for QL = 0.01	
	Filtering	BER = 0 for QL = 0.005 and 0.01	
High-order difference statistics [18]	MP3 compression	SNR greater than 20 for all	SNR values for the attacks on the watermarked signal are very promising. This method provides robustness against MP3 compression, Gaussian noise addition, resampling and requantization attacks
	Gaussian noise	For Gaussian noise with SNR = 20, BER = 0.0001, and with SNR = 15, BER = 0.0015	
	Resampling Requantization	Watermark can be extracted without any errors	
DFT [21]	Insertion attack Removal attack	Detects these types of attacks with very little to zero error rate	Proposed method detects the forgery attacks but makes skew errors during curve approximation

addition, geometric distortions, Gaussian noise, resampling, requantization, insertion attack, and removal attack. The measures used to evaluate the performance of these methods include ODG (objective difference grade), SDG (subjective difference grade), BER (bit error rate), and SNR (signal-to-noise ratio). These methods show promising results against a variety of attacks, but their performance can vary based on the type and intensity of the attack. It is crucial to assess the effectiveness of these methods against various attacks and optimize their parameters to enhance their robustness.

Table 2.7 summarizes the performance metrics of various video authentication techniques that use digital watermarking to protect against different types of attacks. The techniques considered include those based on DWT, QIM, semi-fragile watermarking, and blind extraction schemes. The attacks considered include cropping, rotation, the addition of noise, frame dropping, median filtering, and object insertion/deletion [33–36]. The performance metrics considered include PSNR, SF, NC, and BER. The results show that the different techniques have varying degrees of effectiveness in protecting against different types of attacks, with some techniques performing better than others in terms of the chosen performance metrics. Reference [29] has a

Table 2.7 Analysis of video authentication techniques using performance metrics

Reference	*Attack*	*Performance metrics*
[25]	Gaussian noise salt and pepper cropping rotation	PSNR = 34.97, SF = 0.997 PSNR = 43.48, SF = 0.981 PSNR = 33.3, SF = 0.507 PSNR = 33.23, SF = 0.523
[27]	–	Average PSNR is greater than 50
[28]	–	Average PSNR is greater than 54
[29]	Frame dropping (10–50%), cropping (ratio 10–50), addition of noise (ratio 10–50) median filtering	NC = 0.9599 NC = 0.8609 NC = 0.8641 NC = 0.6303
[30]	–	PSNR = 55.086 at 1080 p, 66.096 at 2160 p
[24]	Gaussian noise salt and pepper cropping rotation	BER = 0.01 (test), 0.0008 (camera2), 0.0409 (video1) BER = 0.021 (test), 0.0018 (camera2), 0.0412 (video1) BER = 0.34 (test), 0.37(video1), 0.3840 (video1) BER = 0.019(test), 0.0009(camera2), 0.03(video1)
[23]	Cropping attack rotation Gaussian Noise Object insertion and deletion	BER = 0.5 (test), 0.5 (camera2), 0.5 (video1). BER = 0.5 (test), 0.5 (camera2), 0.5 (video1). BER = 0.035 (test), 0.04 (camera2), 0.03 (video1). BER = 0.4 (test), 0.4 (camera2)

high similarity factor for frame dropping and cropping attacks, whereas Refs. [24, 25] and Ref. [23] perform much better when subjected to noise addition attacks. It is to be noted that all the PSNR and SNR values are in dB, while the similarity factors have no unit as they represent ratios [37–40].

2.5 CONCLUSION

This chapter presents a comprehensive analysis of different digital watermarking techniques and how they perform against common manipulation attacks. Starting with a brief introduction and classification of digital watermarking techniques, this work focuses on presenting existing image, audio, and video authentication methods. Performance analysis of these techniques is also presented in a detailed manner using several performance metrics used by researchers. Some of the image authentication techniques presented show very promising results by providing high similarity factors and PSNR values.

The technique proposed in Ref. [9] gives the highest PSNR values, and when subjected to several attacks offers admissible robustness. However, there is a scope for improvement in terms of robustness against JPEG compression techniques for methods proposed in Refs. [5, 16]. In general, all image authentication methods discussed have high robustness and great similarity factors.

Regarding audio authentication mechanisms, we have discussed efficient methods which work very well when subjected to several common signal processing attacks. The approach in Ref. [19] gives the highest SNR values, and performs well when subjected to noise addition attacks, but fails for resampling attacks. Reference [17] produces very good results against all types of attacks.

Video authentication techniques also show acceptable results, with PSNRs greater than 50 dB for approaches in Refs. [27, 28, 30]. The methods in Refs. [23–25] do not produce good results when subjected to cropping attacks.

REFERENCES

1. Ingemar Cox et al. "Digital watermarking". Journal of Electronic Imaging 11.3 (2002), pp. 414–414.
2. S. R. Subramanya and B. K. Yi. "Digital signatures". In: IEEE Potentials 25.2 (2006), pp. 5–8.
3. Deepa Kundur and Dimitrios Hatzinakos. "Digital watermarking for telltale tamper proofing and authentication". In: Proceedings of the IEEE 87.7 (1999), pp. 1167–1180.
4. Archana Tiwari, Manisha Sharma, and Raunak Kumar Tamrakar. "Watermarking based image authentication and tamper detection algorithm using vector quantization approach". In: AEU-International Journal of Electronics and Communications 78 (2017), pp. 114–123.
5. Behrouz Bolourian Haghighi, Amir Hossein Taherinia, and Reza Monsefi. "An effective semi-fragile watermarking method for image authentication based on lifting wavelet transform and feed-forward neural network". In: Cognitive Computation 12 (2020), pp. 863–890.

6. Oswaldo Ulises Juarez-Sandoval et al. "Digital image ownership authentication via camouflaged unseen-visible watermarking". In: Multimedia Tools and Applications 77 (2018), pp. 26601–26634.
7. Gerald Schaefer and Michal Stich. "UCID: An uncompressed color image database". In: Storage and Retrieval Methods and Applications for Multimedia 2004. Vol. 5307. SPIE. 2003, pp. 472–480.
8. Duc-Tien Dang-Nguyen et al. "Raise: A raw images dataset for digital image forensics". In: Proceedings of the 6th ACM Multimedia Systems Conference. 2015. New York, pp. 219–224.
9. Thai-Son Nguyen, Chin-Chen Chang, and Xiao-Qian Yang. "A reversible image authentication scheme based on fragile watermarking in discrete wavelet transform domain". In: AEU: International Journal of Electronics and Communications 70.8 (2016), pp. 1055–1061.
10. Abdulaziz Shehab et al. "Secure and robust fragile watermarking scheme for medical images". In: IEEE Access 6 (2018), pp. 10269–10278.
11. Hazem Munawer Al-Otum. "Colour image authentication and recovery using wavelet packets watermarking". In: Circuits, Systems, and Signal Processing 41.6 (2022), pp. 3222–3264.
12. Ertugrul Gul and Serkan Ozturk. "A novel pixel-wise authentication-based self-embedding fragile watermarking method". In: Multimedia Systems 27.3 (2021), pp. 531–545.
13. Jianjing Fu et al. "A watermarking scheme based on rotating vector for image content authentication". In: Soft Computing 24 (2020), pp. 5755–5772.
14. Allan G. Weber. "The USC-SIP Iimage database: Version 5". 2006. http://sipi.usc.edu/database/.
15. ILSVRC 2016 Dataset. 2023. http://image-net.org/challenges/LSVRC/2016/download-images-8r28.php (accessed 3/13/2023).
16. Sajjad Bagheri Baba Ahmadi et al. "An intelligent and blind dual color image watermarking for authentication and copyright protection". In: Applied Intelligence 51 (2021), pp. 1701–1732.
17. Zhenghui Liu et al. "Authentication and recovery algorithm for speech signal based on digital watermarking". In: Signal Processing 123 (2016), pp. 157–166.
18. Xingyuan Liang and Shijun Xiang. "Robust reversible audio watermarking based on high-order difference statistics". In: Signal Processing 173 (2020), p. 107584.
19. Weizhen Jiang, Xionghua Huang, and Yujuan Quan. "Audio watermarking algorithm against synchronization attacks using global characteristics and adaptive frame division". In: Signal Processing 162 (2019), pp. 153–160.
20. Xiao-Chen Yuan, Chi-Man Pun, and C. L. Philip Chen. "Robust Mel-frequency Cepstral coefficients feature detection and dual-tree complex wavelet transform for digital audio watermarking". In: Information Sciences 298 (2015), pp. 159–179.
21. Chang-Mok Park, Devinder Thapa, and Gi-Nam Wang. "Speech authentication system using digital watermarking and pattern recovery". In: Pattern Recognition Letters 28.8 (2007), pp. 931–938.
22. Diego Renza, Camilo Lemus, et al. "Authenticity verification of audio signals based on fragile watermarking for audio forensics". In: Expert Systems with Applications 91 (2018), pp. 211–222.
23. Amal Hammami, Amal Ben Hamida, and Chokri Ben Amar. "Blind semi-fragile watermarking scheme for video authentication in video surveillance context". In: Multimedia Tools and Applications 80 (2021), pp. 7479–7513.
24. Amal Hammami et al. "Regions based semi-fragile watermarking scheme for video authentication." In: Journal of WSCG 28.1–2 (2020), pp. 96–104.

25. Canavoy Narahari Sujatha, Gudipalli Abhishek, and Jeevan Reddy Koya. "Discrete wavelet transform based non-blind video authentication using principal component analysis". In: ICT with Intelligent Applications: Proceedings of ICTIS 2021, Volume 1. Springer. 2022, pp. 425–431.
26. Sonjoy Deb Roy et al. "Hardware implementation of a digital watermarking system for video authentication". In: IEEE Transactions on Circuits and Systems for Video Technology 23.2 (2012), pp. 289–301.
27. Anna Egorova and Victor Fedoseev. "QIM-based semi-fragile watermarking for H. 264 video authentication". In: 2020 8th International Symposium on Digital Forensics and Security (ISDFS). IEEE. 2020, pp. 1–6.
28. Faten H. Mohammed Sediq Al-Kadei and Sohaib Najat Hasan. "Improve a secure blind watermarking technique for digital video". In: Periodicals of Engineering and Natural Sciences 10.2 (2022), pp. 283–291.
29. K. R. Chetan and K. Raghavendra. "DWT based blind digital video watermarking scheme for video authentication". In: International Journal of Computer Applications 4.10 (2010), pp. 19–26.
30. Asem Khmag. "A robust watermarking technique for high-efficiency video coding (HEVC) based on blind extraction scheme". In: SN Computer Science 2.4 (2021), p. 329.
31. Rinaldi Munir et al. "A secure fragile video watermarking algorithm for content authentication based on Arnold Cat Map". In: 2019 4th International Conference on Information Technology (InCIT). IEEE. 2019, pp.32–37.
32. Sharanpreet Kaur et al. "A systematic review of computational image steganography approaches". In: Archives of Computational Methods in Engineering 29 (2022), pp. 1–23.
33. Deb, S., Das, A., & Kar, N. "An applied image cryptosystem on Moore's automaton operating on δ(qk)/𝔽2". In: ACM Transactions on Multimedia Computing, Communications and Applications 20.2 (2023), 1–20.
34. Roy, K. S., Deb, S., & Kalita, H. K. "A novel hybrid authentication protocol utilizing lattice-based cryptography for IoT devices in fog networks". In: Digital Communications and Networks 10.4 (2022), 989–1000. https://doi.org/10.1016/j.dcan.2022.12.003
35. Deb, S., Pal, S., & Bhuyan, B. "NMRMG: Nonlinear multiple-recursive matrix generator design approaches and its randomness analysis". In: Wireless Personal Communications 125.1 (2022), 577–597.
36. Asmitha, P., Rupa, C., Nikitha, S., Hemalatha, J., & Sahu, A. K. Improved multiview biometric object detection for anti spoofing frauds. In: Multimedia Tools and Applications (2024). 1–17. https://doi.org/10.1007/s11042-024-18458-8
37. Sahu, A. K., Umachandran, K., Biradar, V. D., Comfort, O., Sri Vigna Hema, V., Odimegwu, F., & Saifullah, M. A. In: A study on content tampering in multimedia watermarking. In: SN Computer Science 4.3 (2023), 222.
38. Kamil Khudhair, S., Sahu, M., KR, R., & Sahu, A. K. Secure reversible data hiding using block-wise histogram shifting. In: Electronics 12.5 (2023), 1222.
39. Raghunandan, K. R., Dodmane, R., Bhavya, K., Rao, N. K., & Sahu, A. K.. Chaotic-map based encryption for 3D point and 3D mesh fog data in edge computing. In: IEEE Access 11 (2022), 3545–3554.
40. Sahu, A. K.. A logistic map based blind and fragile watermarking for tamper detection and localization in images. In: Journal of Ambient Intelligence and Humanized Computing, 13.8 (2022), 3869–3881.

Chapter 3

Digital image robust information hiding approach

Rupa Jamatia and Bubu Bhuyan

3.1 INTRODUCTION

In the dynamic digital communication landscape and widespread data sharing, secure and effective information-hiding methods have become paramount to safeguard sensitive information from unauthorized access, ensuring confidentiality, integrity, authentication, and privacy. With increasing dependence on digital platforms, the critical importance of implementing resilient methods to conceal information becomes undeniable to mitigate the urgency to address risks linked to data breaches, unauthorized surveillance, and malicious exploitation. The demand for advanced information-hiding strategies directly responds to the inherent vulnerabilities present in digital communication channels, emphasizing the need to strengthen the approaches for secure information exchange, particularly in an era characterized by heightened interconnectivity and reliance on data-driven technologies across diverse domains such as healthcare, finance, transportation, communication, and more. Digital steganography and digital watermarking have distinct purposes; both involve manipulating the host media to embed additional information. They are techniques employed to embed information in digital media for purposes such as authentication, copyright protection, and covert communication. Watermarking intricately embeds digital bits within content to identify the creator or authorized users. Diverging from conventional printed watermarks, their digital counterparts are intentionally imperceptible to viewers. Consequently, the resilience of a digital watermark becomes paramount, requiring the capability to withstand detection, compression, and various operations that may be imposed on a document. Enhancing robustness in digital watermarking and steganography requires the discreet embedding of concealed information to make it capable of withstanding diverse attacks and strengthening the resilience of remaining undetected by unauthorized parties. In this process, information is hidden within the host media and marked with a unique authentication or ownership verification watermark. It provides a way to verify the origin and ownership of the hidden information, deterring unauthorized use and tampering. Robust digital steganography and

DOI: 10.1201/9781032663647-3

watermarking aim to create a secure and resilient method for covert communication, authentication, and ownership protection within the digital domain. This chapter discusses in detail to facilitate a comprehensive understanding; this introduction is thoughtfully organized into distinct subsections, each offering unique insights about digital image steganography. However, the key differences lie in the visibility of the embedded information and the intended purpose. Table 3.1 provides a concise overview of the fundamental key differences between digital steganography and watermarking to enhance a better understanding of their distinctions.

Table 3.1 The key difference between digital watermarking and steganography

	Digital watermarking	*Digital steganography*
Objective	To embed information within digital content to indicate ownership, authorship, or other metadata, which can be a visible or invisible mark that identifies the source or owner of the content	To hide the existence of the embedded information and the communication by concealing messages within the digital media, making it difficult for unauthorized parties to detect the presence of secret data
Visibility	Watermarks are often designed to be perceptible. A watermark is meant to be detected to verify authenticity or ownership. The visibility may vary from subtle alterations to more noticeable markings	Focused on intentionally embedding and hiding data, and detection by unauthorized parties is discouraged. This method aims for imperceptibility, i.e., concealing any alterations and ensuring that the carrier data looks unchanged
Purpose	Used for traceability, copyright protection, authentication, and ownership verification. Commonly employed in the media industry to deter unauthorized use and distribution of digital content	Used for covert communication, hiding sensitive information, and secure data transmission. Applications include secure data transmission, information hiding in images or audio for security purposes, and protecting communication channels
Detection	Detection is often straightforward and intentional. A watermark is usually meant to be detected to establish ownership or verify authenticity	Detection is generally unintentional and challenging. The primary objective is to hide the presence of the embedded information, making it difficult for unauthorized parties to discover or prove its presence
Applications	Digital watermarking includes copyright protection, content authentication, and forensic tracking	Secure messaging, military communications, and privacy-preserving applications
Security criterion	Robustness	Resistance to steganalysis

3.1.1 History of digital watermarking and steganography

Early concepts of digital watermarking started during the 1970s–1990s with the idea of embedding imperceptible patterns into digital images for copyright protection, and later, Thomas C. Cox introduced the term "digital watermark" in a 1987 paper. Initially, it was rooted in the digital media copyright protection field. In the 1990s, digital watermarking gained commercial importance as the Internet and digital media distribution grew. Modern steganography emerged with a focus on hiding digital data. Early techniques involved hiding information within the least significant bits (LSBs)of image files. The research explored various media types for embedding hidden messages, and the efforts in steganalysis—detecting hidden information—also rose with time.

Steganography also found applications in data protection, watermarking, and privacy preservation by employing advanced techniques and methods to conceal information within digital media, making detecting hidden data challenging. The focus is on creating more robust and secure hiding techniques while addressing challenges posed by advanced steganalysis methods.

3.1.2 Hiding information in digital image

In the domain of information/data hiding techniques, digital steganography and watermarking are two methods that have contributed significantly to the security landscape of digital information, ensuring robust data hiding. Steganography and digital watermarking are related concepts within the information security domain, particularly in hiding or embedding information within digital media. Steganography is the practice of concealing information within digital media like images, audio, or video so that it is difficult to detect or decipher, where the sender and receiver want to exchange information without attracting attention. On the other hand, digital watermarking is a technique employed to embed a unique identifier or mark, known as a watermark, into digital content like images, audio, or video. Unlike steganography, the presence of a watermark is usually known to the recipient and is designed to be detectable. Watermarking in digital images is commonly used for copyright protection, authentication, integrity, data protection, ownership, etc. The embedding process in digital steganography, employed to conceal secret data within a cover image, serves as the backbone of the steganographic procedure. In digital steganography, watermarking involves embedding data, usually in the form of a watermark, into digital media like images, audio, video, or text. Various watermarking methods are employed based on the characteristics of the host media and the application's specific requirements based on reversibility, robustness, or steganography domains, as shown in Figure 3.1.

Evaluation of digital steganography and watermarking techniques involves a trade-off between these parameters. Several parameters are used to evaluate the effectiveness and goodness of a steganographic technique, described in Table 3.2.

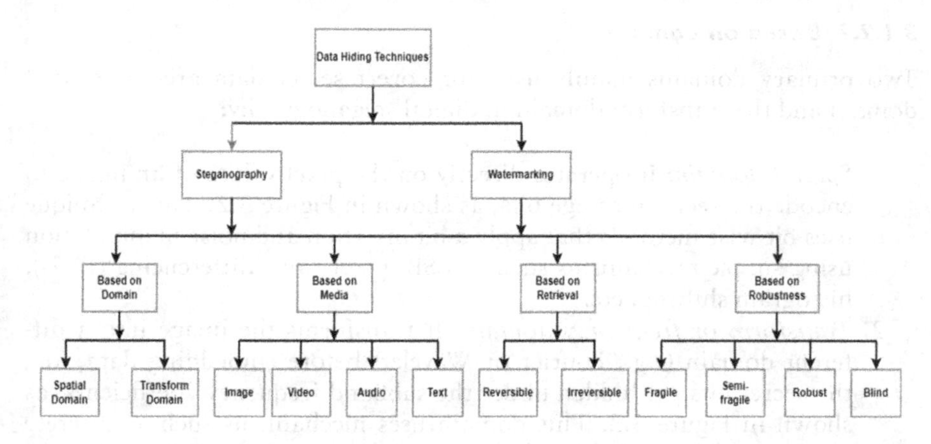

Figure 3.1 Classification of data hiding techniques in digital images.

Table 3.2 Essential characteristics of digital watermarking and steganography

Characteristics	*Digital watermarking*	*Digital steganography*
Invisibility/ imperceptibility	Watermarks should be imperceptible to the human eye or ear. The incorporated data must not compromise the quality of the host media	Like watermarks, steganographic techniques aim to be undetectable by the human senses. The carrier medium (image, audio, video) should appear unchanged after the embedded hidden data
Robustness	Watermarks should withstand everyday signal processing operations and attacks (such as compression, cropping, and filtering) without being destroyed or significantly degraded.	Steganographic methods should be robust against everyday signal processing operations and attacks. The hidden information should survive regular media transformations
Security	Watermarking should provide security to prevent unauthorized removal or alteration. This is crucial for protecting intellectual property and ensuring the authenticity of the content	Steganography should provide security to prevent unauthorized detection of hidden information. The goal is to keep the presence of the embedded data secret
Capacity	The capacity of a watermarking scheme refers to the amount of information that can be embedded without significantly affecting the quality of the host media	Steganography must balance the amount of hidden information with the imperceptibility of the carrier medium. Higher capacity often comes at the cost of increased detectability
Detection and extraction	Reliable methods are required to detect and extract the watermark from the watermarked content. This process should be efficient and accurate	The effectiveness of detection and extraction methods depends on the sophistication of the steganographic techniques employed and the efforts to conceal the hidden information

3.1.2.1 Based on domain

Two primary domains mainly used for covert secret data are the spatial domain and the transform domain in digital steganography:

1. *Spatial domain:* It operates directly on the pixel values of an image to encode the secret message bits, as shown in Figure 3.2. This technique uses bit-wise methods that apply a bit insertion and noise manipulation using simple mechanisms such as LSB, pixel value differencing (PVD), histogram shifting, etc.
2. *Transform or frequency domain:* It transforms the image into a different domain (e.g., Fourier or Wavelet) before embedding data, i.e., the secret bits get hidden under the subband frequency coefficients, as shown in Figure 3.3. This domain uses mechanisms such as discrete cosine transform (DCT), discrete wavelet transform (DWT), integer wavelet transform (IWT), complex wavelet transforms (CWT), spread spectrum, etc.

Table 3.3 provides further details on the differentiation between the spatial domain and transform domain from various perspectives.

3.1.2.2 Based on media

- *Text steganography:* Securing information within text files encompasses various techniques, including modifying the format of existing text, altering words within the text, generating random character sequences, or employing context-free grammar to craft coherent and understandable texts.

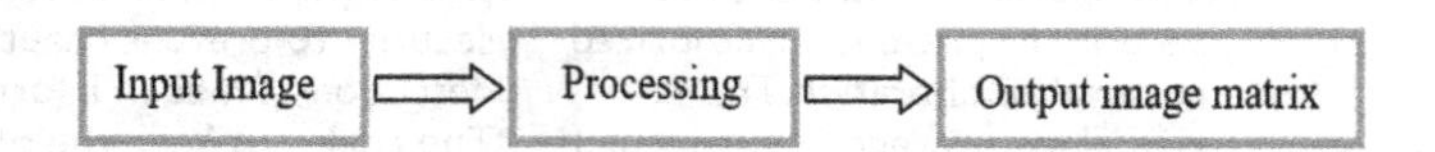

Figure 3.2 Spatial domain steganography.

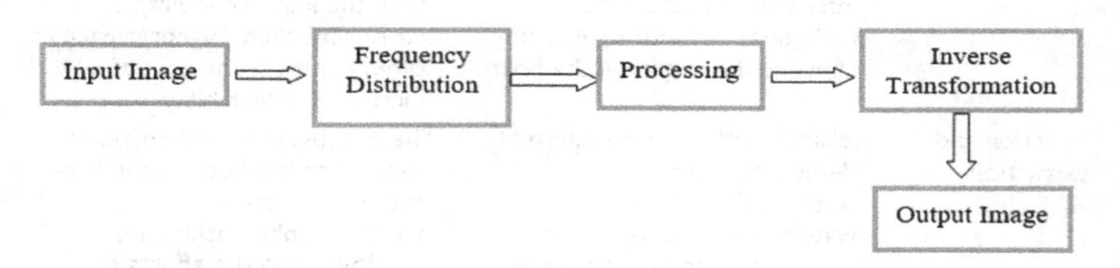

Figure 3.3 Frequency domain steganography.

Table 3.3 Comparison between spatial domain and transform domain techniques

	Spatial domain	*Transform or frequency domain*
Embedding process	It directly deals with images in their natural state, where the pixel values fluctuate concerning the scene	In this context, concealed bits are embedded beneath the subband frequency coefficients. It focuses on the pace of pixel value changes within the spatial domain
Capacity	It provides data embedding higher capacity	Lower capacity as compared to the spatial domain
Complexity	Spatial domain techniques are quite simple	These techniques are complex
Robustness against attacks	Data embedding is more robust to geometrical attacks, such as cropping and down-sampling	Data embedding usually has more robustness to signal processing attacks, such as the addition of noise, compression, and low pass filtering
Perceptual transparency	This method can be designed to be imperceptible to the human eye, preserving the visual quality of the cover media, which may require careful consideration	This technique frequently utilized mathematical transformations and perceptual models to enhance the effectiveness of achieving perceptual transparency components of the signal
Examples	LSB, PVD, etc.	DFT, DCT, etc.

- *Image steganography:* Securing data through covert blending within an image involves embedding information within the visual content. This process often employs techniques such as steganography, where data is subtly integrated into the image's pixel values or other visual elements without perceptible alterations to the human eye.
- *Video steganography:* Data is hidden in digital video format. It offers the benefit of concealing a substantial volume of data within a dynamic flow of images and sounds. It represents a fusion of image steganography and audio steganography.
- *Audio steganography:* Embedding a covert message into an audio signal involves manipulating the binary sequence of the associated audio file. Concealing secret messages in digital sound poses a greater challenge than other methods, such as image steganography, due to the intricate nature of audio data and the complexities associated with imperceptibly altering the binary representation of the sound file.

3.1.2.3 Based on retrieval

Secret data bits or watermarks are used in digital media's steganography and watermarking process. At the recipient's side, these hidden data bits or watermarks at the recipient's side are extracted from the stego-media. The retrieval process can be categorized into two types:

- *Reversible:* The embedding process is designed to be fully reversible, allowing the extraction of the original, unaltered content from the watermarked version without any loss, i.e., the original data can be completely recovered. It is ideal for scenarios where complete preservation of the original content is crucial, mainly used in medical imaging, digital archiving, etc. Prediction error expansion (PEE) and histogram modification are used in this watermarking method.
- *Irreversible:* Irreversible watermarking introduces modifications to the host media in a way that is not intended to be completely reversible, i.e., the recovery may not be perfect, or the original data is modified. It is suitable for scenarios where some loss of fidelity is acceptable in exchange for increased robustness. This method is used in Copyright Protection and Content Authentication.

3.1.2.4 Based on robustness

Regarding digital steganography and watermarking methods, four main types are categorized according to their robustness. They are distinguished by the embedding ability to withstand various signal processing operations, attacks, and distortions while remaining detectable or recoverable. They are as follows:

- *Fragile:* Fragile steganography/watermarking is designed to be highly sensitive to any modifications, making them helpful in detecting tampering or alterations, including intentional and unintentional changes such as modifications to pixel values, compression, or other forms of manipulation. It is commonly used for detecting alterations or tampering in sensitive content, including digital forensics, legal documents, and image authenticity verification.
- *Semi-fragile:* The goal is to find a middle ground between robustness and fragility in steganography or watermarking, allowing the information to survive legitimate operations while detecting malicious tampering such as compression or filtering while being fragile or sensitive to unauthorized modifications.
- *Blind:* It is also known as perceptual steganography or watermarking that allows for the embedding or extracting of information without requiring access to the original during the process. However, it may be challenging or impossible during the extraction phase. It is advantageous in scenarios where the original content is unavailable, such as in distributed systems. Blind steganography or watermarking is commonly used in broadcasting, streaming, or distributed databases.
- *Robust:* It aims to embed data that can withstand attacks and distortions such as compression, filtering, cropping, and geometric distortions.

It is frequently used in copyright protection for digital media, ensuring that ownership information remains intact even after the content transforms. It ensures that the embedded information remains recoverable and intact despite intentional or unintentional modifications.

3.2 RELATED WORK

In the literature review of related work, we have organized the contributions of different authors to digital image steganography and watermarking for robust data hiding.

Choi and Aizawa [1] focus on a digital watermarking technique that utilizes the block correlation of DCT coefficients. They employ the DCT coefficients, commonly used in image and signal processing, to embed a watermark blockwise manner that implies a correlation or relationship between adjacent blocks of DCT coefficients. This technique enhances the robustness and effectiveness of digital watermarking, addressing issues such as resistance to attacks or maintaining the quality of the original content. Shih and Wu [2] explore a combination approach to image watermarking, incorporating techniques in both spatial and frequency domains. Image watermarking is a method used to embed information or a watermark into digital images. The process involves dividing the watermark image into two distinct components—one designated for spatial insertion using LSB and the other for frequency insertion—contingent upon the user's preferences and the significance of the data involved using inverse-DCT. Dividing the watermark into two segments enhances the level of protection twofold. Zhao et al. [3] propose a dual-domain watermarking technique, a DCT–DWT dual-domain algorithm, designed explicitly for safeguarding and compressing cultural heritage imagery that combines image authentication and compression for color images through the implementation of a digital watermarking and data hiding framework. It aims to leverage the strengths of each domain for improved authentication and compression. Their method's advantages are that external watermark information does not need to be transmitted to the receiver using a dual domain. This convenience and enhanced security against eavesdropping come at the cost of increased ambiguity in tamper assessment. Kundur and Hatzinakos [4] introduce FuseMark, an innovative and robust watermarking approach grounded in image fusion principles, designed for applications such as copy protection and resilient tagging. They tackle the intricacies of logo watermarking in still images, employing principles of multi-resolution data fusion to seamlessly embed and extract watermarks. Their method, FuseMark, presently excels by utilizing image fusion tool sets for watermarking, ensuring recoverability for blind detection and automating the determination of maximum overall watermark energy. Peining and Eskicioglu [5] introduced a concept that involves embedding a binary pattern, represented in both the LL and HH bands at the second tier of DWT decomposition. Our approach extends to cover all four bands

(LL, HL, LH, and HH), and we conducted a comparative analysis between first-level and second-level decomposition watermark embedding. Their methods suggest that the first-level decomposition proves advantageous for two reasons: it maximizes the region designated for watermark embedding, and the subsequently extracted watermarks exhibit a more textured and visually superior quality. Ganic et al. [6] propose a circular watermarking concept designed to embed multiple watermarks seamlessly across lower and higher frequencies, with a specific emphasis on making them imperceptible in the DFT domain. The proposed watermarking scheme achieves heightened overall robustness by incorporating multiple frequency bands. They obtained the maximum correlation, robust against cropping, Gamma correction, JPEG compression, and Gaussian noise. Tsui et al. [7] introduce two vector watermarking schemes utilizing complex and quaternion Fourier transforms. A novel aspect of these schemes is the first-time demonstration of embedding watermarks in the frequency domain aligned with the human visual system. The initial approach encodes the chromatic content of a color image into CIE chromaticity coordinates, with the achromatic content represented as CIE tristimulus values. Yellow and blue color watermarks are then seamlessly embedded in the frequency domain of the chromatic channels using a spatial-chromatic discrete Fourier transform. In the second approach, color image components are encoded, and watermarks are seamlessly embedded as vectors in the frequency domain of the channels using the quaternion Fourier transform. A watermark is embedded in the coefficient with positive frequency and spread across color components in the spatial domain to ensure robustness. Invisibility is guaranteed by modifying the coefficient with negative frequency, rendering the cumulative effects imperceptible to the human eye. Parah et al. [8] proposed a system that introduces a novel approach that embedding confidential data within scrambled (encrypted) cover images involves embedding data in the intermediate significant and LSB planes of the encrypted image at specified locations indicated by pseudorandom address space (PAS) and address space direction pointer (ASDP). This process is guided by the principles of scrambling and pseudorandom data embedding. This approach offers a two-layer security mechanism for the covert data and ensures excellent perceptual transparency of the stego images. Vaishnavi and Subashini [9] introduce two methodologies for robust and imperceptible image watermarking within the RGB color space. The initial method involves embedding the grayscale of the watermark into the blue channel elements. In contrast, the second method embeds the blue channel elements of the watermark directly into the host image. Both approaches utilize singular value decomposition (SVD) on the host image's blue channel to embed the watermark within the singular values. Their method improves imperceptibility as the scaling factor increases, while robustness improves with an increasing scaling factor. Najafi [10] propose embedding and detecting watermarks in the DWT domain. The logo watermark elements are directly embedded into the three-level DWT decomposed subbands. Notably, the scheme operates as a completely blind process for both the host image and the watermark. The proposed methodology exhibits desirable attributes such as imperceptibility, blind detection, and resilience against geometrical and non-geometrical attacks. Nazir et al. [11] introduce a blind digital image watermarking technique based on chaotic encryption, suitable for grayscale and color images. Before embedding the watermark, the DCT is applied to the host image. The host image is partitioned into nonoverlapping 8 × 8 blocks before the DCT process, and the watermark bit is covert by altering the difference between the DCT coefficients of adjacent blocks. The proposed scheme exhibits robustness against various image processing operations, including Joint Picture Expert Group (JPEG) compression, sharpening, cropping, and median filtering. The method can be used in e-healthcare and telemedicine to conceal electronic health records in medical images securely. Hannoun et al. [12] introduce a pioneering watermarking scheme operating within the domain of DWT, harnessing the capabilities of a discrete-time chaotic system. The scheme utilizes a modified Henon map as the transmitter and a discrete step-by-step observer as the receiver, ensuring precise synchronization for exact state recovery. The confidential keys within the transmission scheme lie in the transmitter parameters, showcasing both feasibility and robustness in their approach. Dayanand and Ghuli [13] introduce an innovative invisible hybrid watermarking scheme that seamlessly integrates blind and non-blind techniques. The approach employs a blind scheme as the inner watermarking technique, embedding a secret binary image into an inner cover image using DWT. Subsequently, a non-blind watermarking scheme acts as the outer layer, embedding the inner watermarked image into an external cover image through DWT and SVD. This process creates a robust hybrid watermarked image. The hybrid watermarking approach employed in their method demonstrates robustness against various challenges, including rotation, JPEG compression, salt and pepper noise, Gaussian noise, speckle noise, and Poisson noise. Hurrah et al. [14] introduce innovative methods for safeguarding copyright, ensuring data security, and authenticating content within multimedia images. The first scheme addresses copyright protection through a robust watermark that combines DWT and DCT features using an efficient inter-block coefficient differencing algorithm (Scheme I). Scheme II is employed for copyright protection and content authentication by embedding a robust watermark and a fragile logo in the host RGB image. Liu et al. [15] propose an algorithm synergizing DTCWT–DCT, the Henon map, perceptual hashing, and third-party concepts tailored explicitly for medical images with unique requirements. Leveraging zero-watermark technology ensures secure embedding and extraction of watermarks, effectively safeguarding medical images and patients' privacy information. The system demonstrates robustness and resilience against geometric and conventional attacks, emphasizing its steadfastness in geometric attacks. Hussan et al. [16] propose an effective encoding-based watermarking scheme designed for tamper detection and localization. The cover image undergoes division into nonoverlapping 4 × 4 blocks, with each block's arithmetic average (AA) being computed. The AA value is then translated into an

8-bit vector (BV), further augmented to a 16-bit vector through encryption with a watermark. This enhanced bit vector (EBV) is strategically embedded into 4 × 4 blocks, employing Huffman encoding for imperceptibility and DNA encryption for enhanced security. The EBV facilitates tamper detection, while the BV aids in tamper localization. Their method offers lower computing complexity, tamper detection, and localization capabilities, providing an average value of PSNR 51.41 dB. This indicates that the watermarked images maintain high visual quality. The proposed framework is susceptible to various attacks, including copy–paste, copy–move, text addition, and other signal processing attacks.

Additionally, we have compiled and summarized relevant research in digital steganography watermarking techniques from 2002 to 2023, presenting the findings in Tables 3.4–3.6.

Table 3.4 Summarization of the related works of digital steganography watermarking techniques

Authors	*Embedding methods*	*Factors*	*Advantages*	*Applications*
Choi and Aizawa [1]	DCT coefficient	Robustness	Resistance to attacks like JPEG compression and Gaussian noise	Original image quality, illegal copying, and copyright protection
Zhao et al. [3]	DCT–DWT	Security	Improves authentication and compression, tamper resistance	Image authentication
Ganic et al. [6]	DFT and low-and high-frequency bands	Robustness and security	Obtained the maximum correlation, robust against cropping, Gamma correction, JPEG compression, and Gaussian noise	Digital image security
Tsui et al. [7]	SCDFT and QFT	Robustness and imperceptibility	Robust against geometric attacks, Gaussian noise, image enhancement, and better imperceptibility	Copy control and transaction tracking
Parah et al. [8]	LSB, PAS, and address space direction pointer (ASDP)	Security and perceptual transparency	It provides a two-layer security mechanism, high capacity, and perceptual transparency of the stego images	Data security

(Continued)

Table 3.4 Summarization of the related works of digital steganography watermarking techniques *(Continued)*

Authors	*Embedding methods*	*Factors*	*Advantages*	*Applications*
Jia [17]	Blind watermarking and SVD	Robustness and imperceptibility	Robustness against image compression, filtering, cropping, noise adding, blurring, scaling, and sharpening	Copyright protection of digital images
Vaishnavi and Subashini [9]	SVD	Robustness, security	Robust against Gaussian noise, salt-and-noise, motion blur, median filtering, and JPEG compression	Digital image security

Table 3.5 Summarization of the related works

Authors	*Embedding methods*	*Factors*	*Advantages*	*Applications*
Cedillo-Hernandez et al. [10]	DFT	Robustness, watermark quality, payload capacity	Robust against JPEG, sharpening, filtering, and Gaussian noise, geometric attacks. Provide better imperceptibility and PSNR	Medical image management
Laouamer and Tayan [18]	DCT and linear interpolation	Robustness	Provides robustness against filtering, noising, and geometric attacks	Integrity verification, tamper detection, image authentication, and copyright protection
Gaata [19]	DFT	Image quality, robustness	Robust against filtering, blurring, sharpness, and Gamma noise	Copyright protection and authenticity
Haribabu et al. [20]	DWT	Image quality and robustness	Robust against Gaussian noise, salt and pepper noise, speckle noise, and brightness	Copyright protection and owner information

(Continued)

Table 3.5 Summarization of the related works *(Continued)*

Authors	*Embedding methods*	*Factors*	*Advantages*	*Applications*
Jia et al. [21]	DWT and QR decomposition	Robustness and imperceptibility	Robust against compression, cropping, filtering, and noise adding. Provides better imperceptibility	Copyright protection
Najafi [22]	DWT and logo watermark	Robustness and imperceptibility	Provides invisibility, blind detection, and robustness against geometrical and non-geometrical attacks	Copyright protection
Hannoun et al. [12]	DWT and chaotic system	Robustness, security	Secured against statistical attacks	Microcontroller circuits

Table 3.6 Summarization of the related works

Authors	*Embedding methods*	*Factors*	*Advantages*	*Applications*
Singh and Bhatnagar [23]	Integer DCT, non-linear chaotic map, and DSR	Robustness and imperceptibility	Addresses the challenge of false positive detection and ensures resilience against both geometric and non-geometric attacks	Image authentication
Hurrah et al. [14]	DWT, DCT, watermark, and fragile logo	Robustness	Robustness against single/dual/triple attacks, tamper detection, and all signal processing/geometric attacks	Protection of copyright and content authentication
Ambadekar et al. [24]	DWT and encryption	Robustness and imperceptibility	Robust against rotation and salt and pepper noise, and provide better imperceptibility	Copyright protection, content authentication

(Continued)

Table 3.6 Summarization of the related works *(Continued)*

Authors	*Embedding methods*	*Factors*	*Advantages*	*Applications*
Cedillo-Hernandez et al. [25]	DFT and particle swarm optimization	Robustness and imperceptibility	Robustness against geometric distortions	Copyright protection and ownership authentication
Fan et al. [26]	CNN inception V3 and DCT	Robustness and security	Robustness against both conventional and high-intensity geometric attacks	Medical images
Arora et al. [27]	QR code-based visual cryptography	Robustness	Recover the original secret image through the human visual system without any computational processes	Digital image privacy protection
Alomoush et al. [28]	LSB, DCT, and linear modulation	Robustness and imperceptibility	Robust against Gaussian noise, salt and pepper, rotation, cropping, and JPEG compression. Large embedding capacity	Invisible watermarking

3.3 STEGANOGRAPHY FRAMEWORK

As shown in Figure 3.4, the general steganography framework consists of key components such as a secret message, a cover file, a key, and embedding and decoding algorithms. In this framework, the practice commonly

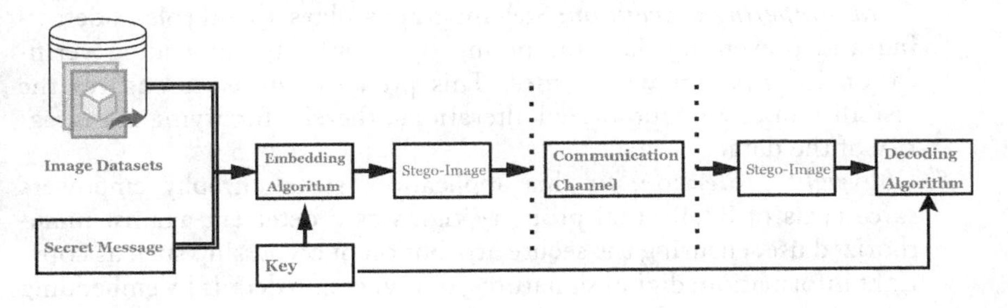

Figure 3.4 General steganography framework.

involves concealing a secret message within a designated transport medium, called the container or carrier. The procedure encompasses encoding at the sender's end to produce the stego-image and subsequent decoding at the receiver's end, unveiling the concealed secret or private information.

3.3.1 Applications of digital steganography

Digital steganography is a technique that serves as a discreet and versatile tool with applications across many domains, as in applications where data confidentiality is a priority. The applications of digital steganography are diverse and impactful, showcasing its adaptability in addressing different needs related to information concealment, secure communication, and data protection within the digital landscape. By embedding copyright information, digital signatures, or ownership details within digital media, steganography helps safeguard the intellectual property rights of creators and acts as a deterrent against unauthorized use, a crucial aspect in fields such as military and defense, forensic investigations, where ensuring the authenticity of digital evidence is essential. The following outlines some of the vital applications employed in digital steganography:

1. *Digital forensics:* In forensic investigations, steganalysis, which involves detecting steganography, is a crucial tool to unveil concealed information within digital media. This systematic approach reveals potential evidence and exposes hidden communications, contributing significantly to the investigative process.
2. *Military and defense communication:* In military and defense communications, steganography plays a pivotal role by concealing sensitive information within seemingly benign digital files. This strategic application enhances communication security by introducing a layer of covert protection to the channels utilized.
3. *Biometric data protection:* Steganography can discreetly conceal biometric data within images or other digital media, ensuring the utmost confidentiality and safeguarding of sensitive information of individuals' biometrics.
4. *Data tampering prevention:* Steganography plays a vital role in detecting and preventing data tampering by seamlessly embedding verification information within files. This proactive measure ensures the visibility of any unauthorized alterations, thereby fortifying the integrity of the data.
5. *Copyright protection:* In this application, steganography empowers safeguards of intellectual property rights as a deterrent against unauthorized use, ensuring the secure attribution of ownership such as copyright information, digital signatures, or ownership details by embedding critical elements within digital media.

3.3.2 Attacks in digital steganography

Digital steganography attacks involve the surreptitious retrieval or alteration of hidden information within digital media, posing threats to data integrity and confidentiality, referred to as steganalysis. The effectiveness of attacks on digital steganography is closely tied to the strength and robustness of the underlying steganographic techniques. Some of the most common attacks in digital steganography include the following:

- *Visual analysis:* This analysis involves the examination and scrutiny of subtle pixel-level changes in digital images or videos to detect the presence of hidden information within digital media, requiring a comprehensive understanding of both the cover medium (the original image or video) and the embedded data either with the computer assistance or naked eye. It is particularly effective when applied to images featuring connected regions of uniform color or areas with saturation at either extreme (0 or 255).
- *Statistical analysis:* This analysis involves the examination of various statistical properties within digital media, enabling the identification of subtle alterations patterns or anomalies that may indicate the presence of hidden information.
- *Histogram analysis:* This involves examining the distribution of pixel intensities or color values in an image by revealing deviations from histograms of the original image and the stego-image.
- *Residual signal (RS) analysis:* This analysis offers a robust and precise method for detecting randomly embedded messages. An image is partitioned into distinct "regular" and "singular" groups based on noise characteristics within each group, such as their mean, variance, or higher-order moments, to detect the presence of hidden information. RS analysis targets a range of the LSB) modification techniques, ensuring thorough and accurate identification of concealed information within the image.
- *Chi-square test:* This test stands out as a renowned and straightforward method for assessing the robustness of a security system against attacks, delivering a highly dependable means of sequentially detecting embedded messages within an image. This examination focuses on steganographic techniques that entail the exchange of pairs of values within pixel gray levels, color schemes, or DCT coefficients.
- *Pairs analysis:* In this analysis, initially, the image is divided into "color cuts," each representing distinct color pairs. This process entails creating a binary sequence for a given color pair, where the first color is assigned a "0," and the second color is assigned a "1". Pair steganalysis typically involves analyzing the differences or residuals between adjacent pixels or samples to detect the presence of hidden information. The objective is to gauge the homogeneity of the LSBs.

3.4 GENERAL MODEL OF DIGITAL WATERMARKING

A digital watermarking model typically consists of several key components and processes designed to embed, detect, and extract watermarks from digital media. The specific algorithms, methods, and parameters used in each component can vary based on the type of digital media (images, audio, video), the application (copyright protection, authentication), and the desired level of robustness and imperceptibility. The essential elements in a typical digital watermarking model consist of the following, as shown in Figure 3.5.

- *Original image* is a digital file often referred to as the cover medium, such as image, video, audio, or text.
- *Watermark* can be a unique identifier, an ownership information, or any other data that should be converted into the original digital image.
- *Watermarked image* is a component to be embedded into the original. During embedding a watermark, a specific algorithm modifies the original data to hide the watermark while minimizing perceptual impact. Later, it will be transmitted along a communication channel.
- *Secret key* ensures the security of the watermarking process and may also be used during the extraction phase.

3.4.1 Applications of digital watermarking

Digital watermarking, like digital steganography, is a robust tool for identifying a document or image's origin, creator, owner, distributor, or authorized consumer. Additionally, it proves invaluable in detecting instances where a document or image has undergone unauthorized distribution or modification. Below are some of the essential applications used in digital watermarking:

1. *Copyright protection:* Digital watermarking enhances copyright protection by embedding invisible identifiers into digital content. An imperceptible watermark is seamlessly embedded into an image, becoming

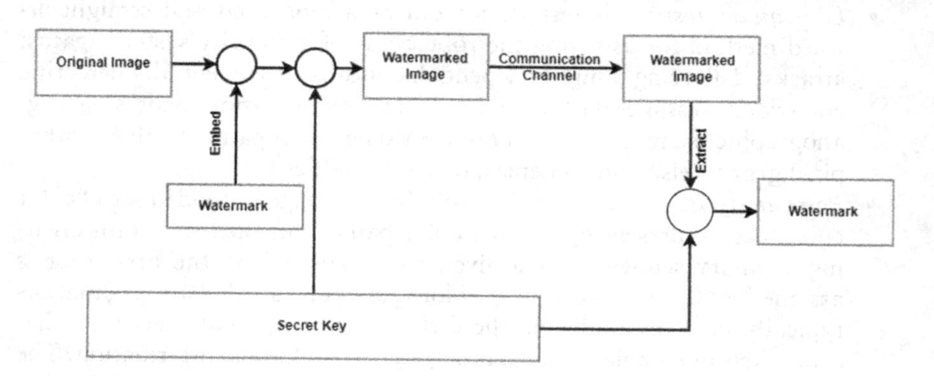

Figure 3.5 General digital watermarking framework.

discernible only through a comparison with the original. Specifically crafted for copyright protection, these watermarks serve the dual purpose of identifying the image's origin and its authorized users. Tamper detection ensures the integrity of the protection, and global standards promote interoperability. Collaboration with digital rights management (DRM) provides a comprehensive defense against copyright infringement across various media formats.

2. *Data authentication:* Data authentication in digital watermarking refers to verifying the integrity and authenticity of digital content by examining the embedded watermark. The objective is to safeguard the content against unauthorized alterations or manipulations, ensuring its integrity and confirming its origin from a reliable and trusted source. Authentication watermarks are mainly used to verify the integrity of documents, images, and multimedia files.
3. *Fingerprinting:* Fingerprinting, also called digital or content fingerprinting, is a form of digital watermarking used for tracking and identifying unique copies of digital content. It involves embedding a unique identifier or "fingerprint" into each copy of the content, making it possible to trace the distribution and usage of that specific copy. Each copy is marked with a unique identifier to trace the origin of unauthorized copies of digital content.
4. *Copy control:* Copy control is an application of digital watermarking to control the number of allowable copies of digital content and prevent unauthorized duplication. Watermarks are embedded in the content to enforce restrictions on copying, ensuring compliance with licensing agreements and copyright protection. Copy control is often integrated into DRM systems, where it helps enforce and manage the usage rights of digital content. It controls access, prevents unauthorized copying, and tracks the distribution of protected content.
5. *Transaction security:* Watermarking is used in secure transactions, such as digital payments and financial transactions, to verify the authenticity of digital documents and ensure that they have not been altered during the transaction process.
6. *Ownership verification:* Ownership verification in digital watermarking is to establish and confirm the rightful ownership of digital content. It is often used when content creators or copyright holders want to assert their ownership rights over a particular piece of digital media.

3.4.2 Attacks in digital watermarking

Digital watermarking is a technique used to embed imperceptible information, known as a watermark, into digital media, such as images, audio, or video, for various purposes, including copyright protection, authentication, and tamper detection. However, like any digital security mechanism, digital watermarks are susceptible to attacks, compromising their integrity or visibility. Some of the common attacks are as follows:

1. *Geometric attacks:* Geometric attacks refer to attacks that aim to distort or manipulate the watermarked content by applying geometric transformations. It can be categorized into several types, each involving a different kind of transformation applied to the watermarked content:
   - *Rotation attacks:* It involves rotating the watermarked content by a certain angle.
   - *Translation attacks:* These involve shifting the position of the watermarked content, either horizontally or vertically, making it difficult for watermark detection algorithms to locate and extract the embedded watermark correctly.
   - *Scaling attacks:* This involves resizing the watermarked content by enlarging or reducing its size.
   - *Shearing attacks:* This involves skewing the watermarked content along one or more axes.
   - *Cropping attacks:* It removes some of the watermarked content.
2. *Signal processing attacks:* These attacks in the context of digital watermarking involve attempts to manipulate or exploit the characteristics of the signal to either remove the watermark or make it undetectable. It targets the digital signal itself, aiming to degrade the quality of the watermark or interfere with its detection.
   - *Noise addition:* In these attacks, attackers may introduce additional noise to the watermarked signal to degrade the quality of the embedded watermark, making it challenging for watermark detection algorithms to distinguish the embedded signal from the added noise.
   - *Blurring attacks:* Blurring attacks, also known as smoothing, involve intentionally applying blurring or smoothing operations that reduce the sharpness of edges and fine details in an image to degrade or remove the embedded watermark. Blurring reduces noise and unwanted variations in pixel values, especially in low-quality or noisy images.
   - *Filtering attacks:* These attacks involve applying filters to the watermarked signal, making it difficult for watermark extraction algorithms to recover the embedded information.
   - *Compression attacks:* These attacks alter a signal's frequency and amplitude characteristics. When a watermarked signal is subjected to compression and subsequent decompression, the embedded watermark may be distorted or completely removed.
   - *Resampling attacks:* These attacks involve changing the sampling rate of the watermarked signal by introducing distortions and altering the temporal characteristics of the signal, affecting the accuracy of watermark extraction.

- *Frequency masking:* These attacks involve adding or emphasizing certain frequencies in the signal to mask the frequencies associated with the embedded watermark to make the watermark less perceptible or harder to detect.

3. *Cryptographic attacks:* These attacks in digital watermarking typically involve an attempt to exploit weaknesses in the cryptographic mechanisms used to secure the watermarking process. It aims to compromise these goals by exploiting the vulnerabilities of encryption, key management, or authentication mechanisms.
   - *Brute force attacks:* Attackers try all possible keys until the correct one is found. The strength of the cryptographic key is crucial in resisting brute force attacks.
   - *Replay attacks:* Capturing and retransmitting valid messages to gain unauthorized access or disrupt the watermarking process.
   - *Man-in-the-middle attacks:* Intercepting and potentially altering communications between different components of the watermarking system.
   - *Forgery attacks:* These attacks can be mainly of two types: key forgery, which aims to attempt to create a fake key or a counterfeit version of a watermarked image by analyzing and mimicking the watermarking technique that can be used to generate valid watermarks. Another one is signature forgery, which aims to create a fake signature to deceive the watermark verification process.
4. *Statistical attacks:* These attacks in digital watermarking involve the analysis and exploitation of statistical properties of the watermarked content or the watermarking process. It leverages statistical information to reveal or manipulate the embedded watermark.
   - *Histogram analysis:* Attackers may examine the histogram of pixel values in the watermarked image to identify patterns or anomalies that could indicate the presence of a watermark.
   - *Correlation analysis:* Attackers may calculate correlations between different regions of the watermarked content to identify areas with higher or lower watermark strength.
   - *Feature extraction:* Extraction of specific features, such as texture or color features, to identify patterns introduced by the watermarking process. Feature-based attacks exploit the characteristics of the embedded watermark.

3.5 EVALUATION METRICS IN DIGITAL STEGANOGRAPHY AND WATERMARKING SYSTEMS

The evaluation metrics are pivotal in assessing the effectiveness and performance of digital steganography and watermarking systems. These metrics are quantitative measures to gauge various aspects, ensuring a comprehensive understanding of the system's capabilities. Parameters such as imperceptibility, robustness against attacks, payload capacity, and computational efficiency are commonly used to evaluate digital steganography systems. On the other hand, watermarking systems are often scrutinized based on criteria like fidelity, resistance to tampering, and the ability to withstand signal processing operations. Some metrics that aim to design, benchmark, and improve these covert communication and information protection methods are as follows.

3.5.1 Measuring noise and distortion resistance

- Evaluating noise and distortion resistance involves applying controlled levels of noise or distortion to the stego-image and then assessing the quality of the hidden data retrieval.
  - *Peak signal-to-noise ratio (PSNR):* PSNR is a widely used metric to measure the quality of the recovered image compared to the original image. A higher PSNR value indicates better resistance to noise and distortion.

$$PSNR = \left(20\log_{10}(R/\sqrt{MSE}\,\right) \tag{3.1}$$

  where R is the maximum fluctuation in the input image data type, 255 for an 8-bit grayscale image.
  - *Mean square error (MSE):* It is used to quantify the visual or auditory quality of the watermarked content compared to the original [29, 30].

$$MSE = \left(\frac{1}{n+m}\sum_{i=0}^{m-1}\sum_{j=0}^{n-1}\left(O(i,j)-D(i,j)\right)^2\right) \tag{3.2}$$

  where O represents the matrix data of the original image, D represents the matrix of embedded message stego-image, m represents the number of rows of pixels, i represents the index of that row of the image, n represents the number of columns of pixels, and j represents the index of that column of the image.
  - *Structural similarity index (SSIM):* SSIM measures the structural similarity between two images. It considers luminance, contrast, and structure, making it suitable for evaluating changes in image quality due to noise and distortion [31, 32].

$$SSIM(x,y) = \frac{(2\mu_x\mu_y + C_1) + (2\sigma_{xy} + C_2)}{(\mu_x^2 + \mu_y^2 + C_1)(\sigma_x^2 + \sigma_y^2 + C_2)} \tag{3.3}$$

  where μ_x and μ_y are the main intensity values of images x and y. $\sigma_x{}^2$ is the variance of x and $\sigma_y{}^2$ is the variance of y. C_1 and C_2 are the two stabilizing parameters, L is the dynamic range of pixel values 2^n bits per pixel-1, and the contents $C_1 = (k_1L)^2$ and $C_1 = (k_2L)^2$ where $k_1 = 0.01$ and $k_2 = 0.03$.

- *Bit error rate (BER):* BER measures the accuracy of recovered bits of hidden data compared to the original bits. A lower BER indicates better resistance to noise-induced errors and better performance, which means fewer errors in the received signal. The BER is the ratio of the number of bits received in error to the total number of bits transmitted. BER is a crucial metric in assessing the reliability and performance of digital communication systems.

$$BER\ (dB) = -10 * \log_{10}\left(\frac{Number\ of\ Bits\ Received\ in\ Error}{Total\ Number\ of\ Bit\ Transmitted}\right) \tag{3.4}$$

- *Normalized cross-correlation (NCC):* It is a metric commonly used in signal processing and image analysis to measure the similarity between two signals or images. NCC is often used to quantify the similarity between the original and watermarked signals after an attack or distortion in evaluating watermarking systems [33, 34].

$$NCC\ (x,y) = \frac{\sum_{i=1}^{n}(x_i - \overline{x})(y_i - \overline{y})}{\sqrt{\sum_{i=1}^{n}(x_i - \overline{x})^2 \cdot \sum_{i=1}^{n}(y_i - \overline{y})^2}} \tag{3.5}$$

where x and y are the signals or vectors being compared. n is the length of the signals. x^- and y^- are the means of the signals x and y, respectively. The resulting NCC value will be between –1 to 1, where 1 indicates a perfect match, –1 indicates a perfect mismatch, and 0 indicates no correlation.

- *SNR:* It is a measure used to quantify a signal's level relative to the background noise level. It is commonly used in various fields, including signal processing, telecommunications, and image processing, to assess the quality of a signal by comparing the strength of the desired signal to the level of unwanted noise. The SNR is expressed in decibels (dB), where higher SNR indicates better quality and resistance to noise and distortions [35, 36].

$$SNR\ (dB) = 10 * \log_{10}\left(\frac{\text{Signal Power}}{\text{Noise Power}}\right) \tag{3.6}$$

where signal power is the power of the signal. Noise power is the power of the noise.

- *Correlations coefficient (CC):* It is often used to quantify the relationship between the original signal (cover image) and the watermarked signal (cover image with embedded watermark).

The correlation coefficient helps measure how well the watermark is hidden or embedded in the cover image. The correlation coefficient is a statistical measure that ranges from –1 to 1, where 1 indicates a perfect positive correlation (the two signals are perfectly correlated), –1 indicates a perfect negative correlation (the two signals are perfectly inversely correlated), and 0 indicates no correlation. A high correlation coefficient between the cover and watermarked images in digital watermarking suggests that the watermark is well hidden. In contrast, a low correlation coefficient may indicate that the watermark is detectable or has affected the cover image significantly. This formula combines the key components of covariance and standard deviations to measure the correlation between the two signals, indicating how well the watermark is embedded in the cover image in the context of digital steganography and watermarking.

$$CC(Y, Y) = \frac{\sum_{i=1}^{n}(X_i - \bar{X})(Y_i - \bar{Y})}{\sqrt{\sum_{i=1}^{n}(X_i - \bar{X})^2 \cdot \sum_{i=1}^{n}(Y_i - \bar{Y})^2}} \tag{3.7}$$

where cov(X, Y) is the covariance between signals X and Y. X^- and Y^- are the means of signals X and Y, respectively. σX is the standard deviation of the signal X. σY is the standard deviation of the signal Y. n is the number of data points.

- *Payload or embedding capacity (EC):* Payload capacity refers to measuring the volume of information within the cover image, i.e., the maximum amount of information that can be successfully embedded, transmitted, and extracted by the steganography process. The communication process mostly depends on the maximum payload capacity, measured in bits per pixel (BPP).

$$BPP = \frac{Number\ of\ secret\ bits\ embedded}{Total\ number\ of\ pixels} \tag{3.8}$$

where the total number of pixels is (image width × image length).

- *Computation complexity:* Computation complexity, an essential metric in digital steganography and watermarking, gauges the computational resources and processing power necessary for successfully implementing the steganography and watermarking techniques within a digital system. Lower computation complexity is desirable as it indicates more efficient and faster processing, making the watermarking method practical for various applications. It is essential to consider factors such as the number of computations per pixel, the number of iterations, and any other relevant parameters to derive a more detailed computational complexity formula for specific digital steganography and watermarking algorithms.

- *Accuracy:* In digital steganography and watermarking, accuracy gauges the effectiveness of concealing or extracting information without introducing perceptible distortions. It reflects the system's ability to preserve integrity while minimizing detectability by unauthorized parties. Metrics such as true positives (TP), true negatives (TN), false positives (FP), and false negatives (FN) are commonly used to evaluate the performance of classification systems to help assess the accuracy and effectiveness of the classification or detection process in digital steganography and watermarking. In steganography, accuracy measures how well-hidden information is embedded, ensuring covert success. High accuracy in steganography ensures that concealed information remains intact and undetected, facilitating the success of covert communication or data hiding. TP measures how well the system identifies the presence of hidden information, while TN measures its ability to correctly identify cases without hidden information. FP and FN indicate errors in detection, helping assess the system's overall performance. In watermarking, accuracy ensures precise extraction even after potential distortions. Accurate watermarking ensures reliable recovery of the original information encoded in the watermark, even after potential distortions or attacks on the watermarked media. TP measures how well the system correctly extracts the watermark, while TN measures its ability to identify cases without a watermark correctly. FP and FN indicate errors in extraction, helping assess the system's overall accuracy. In broader testing, accuracy computes correctly classified samples' ratios, indicating the classifier's efficacy in distinguishing between classes based on features like residuals extracted from images. Formula to calculate accuracy:

$$Accuracy = \frac{(TP + TN)}{(TP + TN + FP + FN)} \tag{3.9}$$

where TP is the accurate identification number of suspicious files with hidden information; TN is the accurate identification of innocent files without hidden information; FP is the count of innocent files mistakenly identified as containing hidden information (false alarms); and FN is the count of suspicious files mistakenly identified as devoid of hidden information.

3.6 CONCLUSIONS

As technology advances, the landscape of digital image steganography undergoes dynamic transformations as new methods are developed, offering opportunities such as secure communication, authentication, data, and copyright protection mechanisms in digital systems. With the development, challenges such as security against advanced attacks and vulnerabilities also increase. Achieving an effective solution in this field requires carefully balancing robustness, security, imperceptibility, and a higher bit embedding rate. The trade-off involves finding the proper equilibrium among these critical factors. To improve the efficiency and effectiveness of digital steganography and watermarking processes, two specific aspects need attention: one is selecting an appropriate cover image to embed the information, and another is the algorithm to embed information within the cover image, making the process more secure, efficient, or capable of handling various attacks. The chapter emphasizes the importance of robust information hiding in digital images, delving into a comprehensive discussion of two fundamental techniques: steganography and watermarking. These techniques are explored for their effectiveness in achieving robust data hiding. It underscores the significance of balancing payload capacity, robustness, and image quality to contribute to secure and effective hidden information transmission. Imperceptibility in the robust data-hiding aspect is pivotal when embedding information, as it avoids causing noticeable changes and maintains the covert nature of the process. The bit embedding rate or capacity represents the amount of concealed information, and a higher rate allows for more data to be hidden. Highlighting the pivotal role of robustness, it's crucial to withstand ubiquitous operations like compression and filtering, ensuring the output seamlessly preserves the original embedded information without discernible changes. However, optimizing these factors simultaneously is challenging, as enhancing robustness may impact imperceptibility, and increasing the embedding rate could compromise security. Furthermore, we discuss different attacks, applications, and existing techniques for robust data hiding in digital steganography and watermarking techniques and how security involves resisting detection. The future work in digital steganography and watermarking could be developing and exploring more hybrid methods that combine the strengths of both steganography and watermarking to achieve an optimized balance between robustness, security, imperceptibility, and bit embedding rate, machine learning integration to enhance the adaptability and intelligence of information hiding systems, and quantum steganography and watermarking that leverage the unique properties of quantum systems, potentially improving security and robustness.

REFERENCES

1. Choi, Yoonki, and Kiyoharu Aizawa. "Digital watermarking technique using block correlation of DCT coefficients." Electronics and Communications in Japan (Part II: Electronics) 85.9 (2002): 23–31.
2. Shih, Frank Y., and Scott YT Wu. "Combinational image watermarking in the spatial and frequency domains." Pattern Recognition 36.4 (2003): 969–975.
3. Zhao, Yang, Patrizio Campisi, and Deepa Kundur. "Dual domain watermarking for authentication and compression of cultural heritage images." IEEE Transactions on Image Processing 13.3 (2004): 430–448.
4. Kundur, Deepa, and Dimitrios Hatzinakos. "Toward robust logo watermarking using multiresolution image fusion principles." IEEE Transactions on Multimedia 6.1 (2004): 185–198.

5. Tao, Peining, and Ahmet M. Eskicioglu. "A robust multiple watermarking scheme in the discrete wavelet transform domain." Internet Multimedia Management Systems V. Vol. 5601. SPIE, 2004.
6. Ganic, Emir, Scott D. Dexter, and Ahmet M. Eskicioglu. "Embedding multiple watermarks in the DFT domain using low-and high-frequency bands." Security, Steganography, and Watermarking of Multimedia Contents VII. Vol. 5681. SPIE, 2005.
7. Tsui, Tsz Kin, Xiao-Ping Zhang, and Dimitrios Androutsos. "Color image watermarking using multidimensional Fourier transforms." IEEE Transactions on Information Forensics and Security 3.1 (2008): 16–28.
8. Parah, Shabir A., et al. "Data hiding in scrambled images: A new double layer security data hiding technique." Computers & Electrical Engineering 40.1 (2014): 70–82.
9. Vaishnavi, D., and T. S. Subashini. "Robust and invisible image watermarking in RGB color space using SVD." Procedia Computer Science 46 (2015): 1770–1777.
10. Cedillo-Hernandez, Manuel, et al. "Robust watermarking method in DFT domain for effective management of medical imaging." Signal, Image and Video Processing 9 (2015): 1163–1178.
11. Loan, Nazir A., et al. "Secure and robust digital image watermarking using coefficient differencing and chaotic encryption." IEEE Access 6 (2018): 19876–19897.
12. Hannoun, Katia, et al. "A novel DWT domain watermarking scheme based on a discrete time chaotic system." IFAC-Papers OnLine 51.33 (2018): 50–55.
13. Savakar, Dayanand G., and Anand Ghuli. "Robust invisible digital image watermarking using hybrid scheme." Arabian Journal for Science and Engineering 44.4 (2019): 3995–4008.
14. Hurrah, Nasir N., et al. "Dual watermarking framework for privacy protection and content authentication of multimedia." Future Generation Computer Systems 94 (2019): 654–673.
15. Liu, Jing, et al. "A robust multi-watermarking algorithm for medical images based on DTCWT-DCT and Henon map." Applied Sciences 9.4 (2019): 700.
16. Hussan, Muzamil, et al. "An efficient encoding based watermarking technique for tamper detection and localization." Multimedia Tools and Applications 82 (2023): 1–23.
17. Jia, Shao-li. "A novel blind color images watermarking based on SVD." Optik 125.12 (2014): 2868–2874.
18. Laouamer, Lamri, and Omar Tayan. "A semi-blind robust DCT watermarking approach for sensitive text images." Arabian Journal for Science and Engineering 40 (2015): 1097–1109.
19. Gaata, Methaq T. "An efficient image watermarking approach based on Fourier transform." International Journal of Computer Applications 136.9 (2016): 8–11.
20. Haribabu, Maruturi, Ch Hima Bindu, and K. Veera Swamy. "A secure & invisible image watermarking scheme based on wavelet transform in HSI color space." Procedia Computer Science 93 (2016): 462–468.
21. Jia, Shaoli, Qingpo Zhou, and Hong Zhou. "A novel color image watermarking scheme based on DWT and QR decomposition." Journal of Applied Science and Engineering 20.2 (2017): 193–200.
22. Najafi, E. J. M. S. "A robust embedding and blind extraction of image watermarking based on discrete wavelet transform." Mathematical Sciences 11 (2017): 307–318.
23. Singh, Satendra Pal, and Gaurav Bhatnagar. "A new robust watermarking system in integer DCT domain." Journal of Visual Communication and Image Representation 53 (2018): 86–101.

24. Ambadekar, Sarita P., Jayshree Jain, and Jayshree Khanapuri. "Digital image watermarking through encryption and DWT for copyright protection." Recent Trends in Signal and Image Processing: ISSIP 2017. Springer Singapore, 2019.
25. Cedillo-Hernandez, Manuel, Antonio Cedillo-Hernandez, and Francisco J. Garcia-Ugalde. "Improving DFT-based image watermarking using particle swarm optimization algorithm." Mathematics 9.15 (2021): 1795.
26. Fan, Yu, et al. "A multi-watermarking algorithm for medical images using inception V3 and DCT." Computers, Materials & Continua 74.1 (2023).
27. Arora, Akanksha, Hitendra Garg, and Shivendra Shivani. "Privacy protection of digital images using watermarking and QR code-based visual cryptography." Advances in Multimedia 2023 (2023).
28. Alomoush, Waleed, et al. "Digital image watermarking using discrete cosine transformation based linear modulation." Journal of Cloud Computing 12.1 (2023): 1–17.
29. Asmitha, P., Rupa, C., Nikitha, S., Hemalatha, J., and Sahu, A. K. "Improved multiview biometric object detection for anti spoofing frauds." Multimedia Tools and Applications (2024): 1–17. https://doi.org/10.1007/s11042-024-18458-8
30. Sahu, A. K., Umachandran, K., Biradar, V. D., Comfort, O., Sri Vigna Hema, V., Odimegwu, F., and Saifullah, M. A. "A study on content tampering in multimedia watermarking." SN Computer Science 4.3 (2023): 222.
31. Kamil Khudhair, S., Sahu, M., Raghunandan, K. R., and Sahu, A. K. "Secure reversible data hiding using block-wise histogram shifting." Electronics 12.5 (2023): 1222.
32. Raghunandan, K. R., Dodmane, R., Bhavya, K., Rao, N. K., and Sahu, A. K. "Chaotic-map based encryption for 3D point and 3D mesh fog data in edge computing." IEEE Access 11 (2022): 3545–3554.
33. Sahu, A. K. "A logistic map based blind and fragile watermarking for tamper detection and localization in images." Journal of Ambient Intelligence and Humanized Computing 13.8 (2022), 3869–3881.
34. Deb, S., Das, A., and Kar, N. "An applied image cryptosystem on Moore's automaton operating on δ(qk)/𝔽2." ACM Transactions on Multimedia Computing, Communications and Applications, 20.2 (2023), 1–20.
35. Roy, K. S., Deb, S., and Kalita, H. K. (2022). "A novel hybrid authentication protocol utilizing lattice-based cryptography for IoT devices in fog networks." Digital Communications and Networks 10.1 (2023). https://doi.org/10.1016/j.dcan.2022.12.003
36. Deb, S., Pal, S., and Bhuyan, B. "NMRMG: Nonlinear multiple-recursive matrix generator design approaches and its randomness analysis." Wireless Personal Communications, 125.1 (2022): 577–597.

Chapter 4

Safeguarding multimedia in the quantum age

Swarna Panthi and Bubu Bhuyan

4.1 INTRODUCTION

The influence of multimedia in our daily lives is unmistakable, seamlessly woven into our routines, encompassing communication and entertainment consumption. The essential requirements for integrity, confidentiality, non-repudiation, and authenticity in both transmitting and storage of multimedia underscore the significance of cryptography, positioning it as a pivotal discipline within the realm of information technology. The utilization of cryptography for the security of multimedia is essential to protect sensitive information, maintain content integrity, and instill trust in digital communication platforms. Its relevance spans diverse domains, playing a crucial role in establishing a secure and trustworthy digital ecosystem [1]. Cryptography is a specialized field dedicated exclusively to ensuring the security of multimedia data. Countless researchers in this domain strive to uphold the privacy and integrity of data by employing a variety of security algorithms. Two primary types of cryptographic systems are currently in use: symmetric and asymmetric. In symmetric cryptography, only one key is used by the sender as well as the receiver for encrypting and decrypting the data. In contrast, asymmetric cryptography, also known as public key cryptography, employs distinct public–private key pairs for encryption and decryption. In this scenario, each communicating party possesses its own set of private and public keys. In recent years, quantum computing has gained significant attention as a research focus, driven by the rapid pace of technological advancements. This trend has brought quantum computers into the spotlight, harnessing quantum physical phenomena to execute calculations. Quantum computers pose a substantial threat to existing public key algorithms and symmetric key algorithms. The consequence of technological progress toward quantum computing is the imminent breakdown of current secure public key algorithms [2]. As a result, any protocols reliant on either key exchange or digital signatures face the risk of compromise, leading to insecure network communications. The National Institute of Standards and Technology (NIST) released the following information in August 2016 [3]: The advent of practical quantum computing will break all commonly used public key cryptographic

DOI: 10.1201/9781032663647-4

algorithms. In response, NIST is researching cryptographic algorithms for the public key-based key agreement and digital signatures that are not susceptible to cryptanalysis by quantum algorithms. Following this news release, numerous public key candidates have undergone extensive scrutiny in recent years, with a focus on their resilience against quantum-based attacks. These cryptographic algorithms, designed to withstand quantum threats, are collectively known as post-quantum cryptography (PQC) [4]. The primary objective of PQC is to create cryptographic systems that remain secure in the face of both traditional and quantum computers, while also ensuring compatibility with existing communication protocols. PQC encompasses five main types: lattice-based, multivariate, code-based, hash-based (HB), and isogeny cryptography. In this chapter, our focus will be on examining the influence of quantum algorithms on existing public key cryptosystems (PKC). Subsequently, we will delve into potential future trajectories for cryptography, exploring the current state of standardization processes for PQC algorithms. Additionally, we will provide a concise overview of the progress made in the development of quantum computers till now and the prospects for researchers to work in the field of PQC.

4.2 CLASSICAL CRYPTOGRAPHY

In this section, we will give a concise introduction to both symmetric and asymmetric cryptography. The security underpinning these cryptographic algorithms hinges on some strong mathematical problems, such as the problem of integer factorization, etc., which will be elucidated in this section.

4.2.1 Symmetric cryptography

This cryptography is a cryptographic approach in which both the encryption and decryption algorithms operate using a single key (Figure 4.1). With sufficiently long and random key lengths, current computational capabilities are unable to feasibly execute brute force attacks against these schemes.

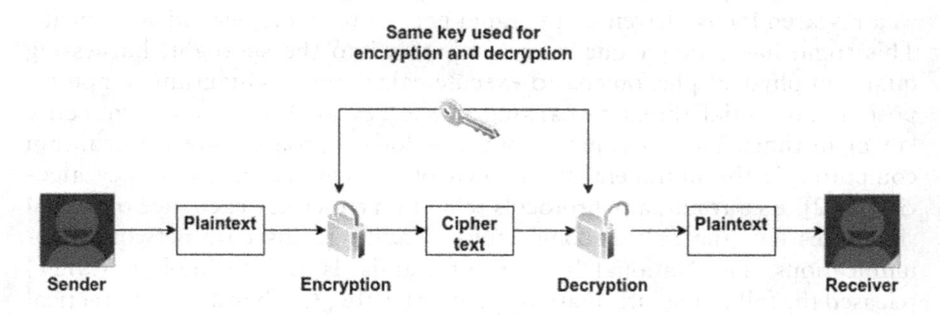

Figure 4.1 Example of working of symmetric key cryptography.

Nonetheless, securely transmitting the key from the sender to the receiver poses a considerable challenge in the effective implementation of symmetric cryptography schemes. This challenge arises due to the reliance on one key for encryption as well as for decryption. Common symmetric algorithms encompass the advanced encryption standard (AES) and the data encryption standard (3DES).

4.2.2 Asymmetric cryptography

In these types of cryptography algorithms, two distinct keys are used: a public key, which is shared between the sender and the receiver, and this key is used by the sender to encrypt the data; and a private key is kept secret by the receiver and the receiver use this key to decrypt the sent data (Figure 4.2). Each participant in the communication process possesses its own set of private and public keys. Asymmetric cryptography finds applications in encryption, key exchange, and digital signatures. In a public key encryption (PKE) algorithm, anyone holding the public key can encrypt a message, producing a ciphertext. However, decryption of the ciphertext to unveil the original message can only be performed by those possessing the corresponding private key. Asymmetric cryptography encompasses various algorithms, including Rivest Shamir Adleman (RSA), the digital signature algorithm (DSA), elliptic curve cryptography (ECC), and others.

4.2.3 Integer factorization problem

The security of RSA lies in the proven difficulty of factoring a large integer. Specifically, RSA leverages the challenge of factoring bi-prime numbers. In the RSA algorithm, the public key comprises two numbers, with one being the product of two large prime numbers. Similarly, the private key is derived from the same pair of prime numbers. However, the factorization problem could be potentially solved if a rapid factorizing algorithm is introduced or if there is a substantial increase in computational power. The latter scenario becomes feasible through the application of quantum mechanics in computing, known as quantum computers [5].

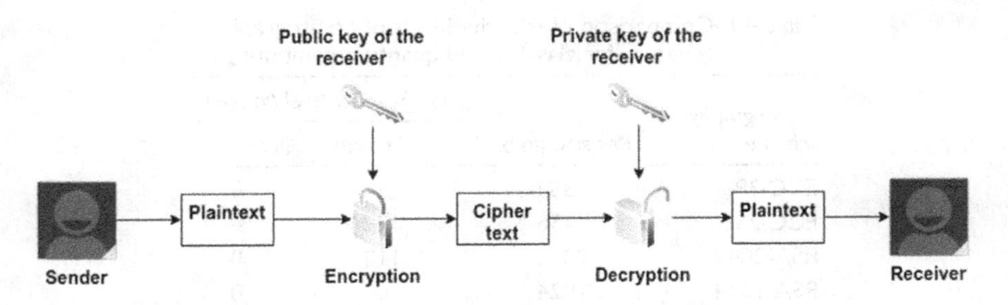

Figure 4.2 Example of working of symmetric key cryptography.

4.2.4 Discrete logarithm problem (DLP)

Numerous public key cryptography, like Diffie–Hellman, ECC, etc., rely on the assumption that finding an efficient solution to the DLP is challenging [6]. Discrete logarithms refer to logarithms defined within the cyclic group of integers modulo n. The problem of a discrete logarithm is articulated as follows: in a group g with a generator G and an element x of g, the task is to determine the logarithm to the base of x within the group g. The computation of the DLP becomes exceedingly challenging when the parameters involved are sufficiently large. However, this paradigm undergoes a shift with the advent of quantum computers, as explained in the subsequent sections.

4.3 IMPACT OF QUANTUM ALGORITHMS ON CLASSICAL CRYPTOGRAPHY

The concept of a quantum computer is deeply rooted in quantum physics, the study of the behavior of matter and energy at the sub-atomic level. In the 1980s, scientists applied the principles of quantum physics and its associated mathematics to conceptualize computers capable of executing specific computations at an exceptionally high rate compared to classical computers. Unlike the binary bits in classical computers, which can represent either ones or zeros, qubits, the fundamental units of quantum information, have the capacity to hold much richer information. This unique feature allows us to leverage qubits to more efficiently calculate the boundaries around interesting events, employing a multitude of dimensions that surpass what can be visualized or handled by a classical computer. They excel at solving problems that are practically insurmountable for classical computers within a limited time frame. Quantum computers have the potential to compute results that are currently beyond the reach of even classical supercomputers. Hence, PKCs that exhibit security on classical computers may become insecure when subjected to quantum computers. While certain variants of symmetric key algorithms remain secure under quantum attacks, ongoing research is delving into their security implications. Table 4.1 provides a comparative overview of classical and quantum security levels for widely used cryptographic schemes.

Table 4.1 Comparison of security levels of cryptography schemes for classical and quantum computing

Cryptography scheme	*Key size (in bits)*	*Security level (in bits)*	
		Classical	*Quantum*
ECC-384	384	256	0
ECC-256	256	128	0
RSA-2048	2048	112	0
RSA-1024	1024	80	0
AES-256	256	256	**128**
AES-128	128	128	**64**

The applications of quantum computers span a wide range of fields, such as simulation of chemical reactions, efficient systems for navigation, information processing, and more [7].

Quantum computers, by harnessing the principles of superposition and entanglement, introduce a paradigm shift in information processing capabilities [8].

Entanglement: Entanglement is a quantum mechanical phenomenon that establishes a correlation in the behavior of two distinct entities. In the context of quantum computing, when two qubits are entangled, any changes made to one qubit instantaneously affect the other. This interdependence is a unique feature of quantum systems. Quantum algorithms strategically exploit these entangled relationships to derive solutions to complex problems, unlocking the potential for parallel computation and enhancing the efficiency of quantum computations.

Quantum parallelism: In traditional computers, bits can exist in one of two states: 0 or 1. Quantum computers, on the other hand, utilize quantum bits, or qubits, which can exist not only in the 0 or 1 state but also in a superposition of both states simultaneously. The distinctive phenomenon known as superposition enables quantum computers to concurrently process information, markedly enhancing their computational capacity in comparison to classical computers.

4.3.1 Shor's algorithm

Shor introduced a quantum algorithm capable of solving two fundamental challenges that underpin the security of contemporary cryptosystems: the problem of prime factorization and the discrete logarithm [9]. Classical solutions to these problems exhibit exponential time complexity. Shor's innovative approach involves utilizing quantum Fourier transformation and period finding to construct probabilistic algorithms that operate in polynomial time.

Shor's algorithm is probabilistic, meaning it doesn't consistently factorize the number in the initial attempt. However, with successive runs, the probability of obtaining a factor increases. This probabilistic nature does not diminish the algorithm's significance, as it still outperforms classical solutions for certain problems with its polynomial time complexity.

Indeed, if a quantum computer with a sufficiently large number of qubits comes into existence, Shor's algorithm has the potential to make vulnerable widely used PKC schemes. This includes but is not limited to the RSA scheme, the finite field Diffie–Hellman key exchange, the elliptic curve Diffie–Hellman key exchange, and more.

Shor's algorithm, by efficiently solving these mathematical problems, would compromise the security of these cryptographic systems, rendering them vulnerable to quantum attacks.

The working of Shor's algorithm is explained below:

1. Choose an x, where $1 < x < N$ and $g(x,N) = 1$, where $N = p \times q$ and p,q.
2. Let r be a random integer and be the period of modular exponentiation of x^a mod N.

3. With a good approximation of r, the $g(x^{r/2} - 1, N)$ and $g(x^{r/2} + 1, N)$ has a good chance of containing p and/or q.

Here g is an abbreviation of greatest common divisor.

4.3.2 Grover's algorithm

Grover's algorithm is a quantum search algorithm designed to search through an unsorted database in $(O(n))$ time complexity [10]. In classical computers, searching an unordered database is an $(O(n))$ problem. Grover's algorithm is probabilistic and can be iterated multiple times to ensure a comprehensive search. Each iteration of Grover's algorithm involves a search, and repeating iterations enhances the probability of locating the desired element in the database. Beyond database search applications, Grover's algorithm can be utilized for tasks such as finding the mean and median of data or determining the inverse value of a function. Quantum cryptanalysts may employ these tools for breaking algorithms, including searching for keys in symmetric cryptography algorithms.

4.3.3 Simon algorithm

Simon's algorithm [11] is a quantum circuit designed to determine if a given function is either one-to-one (1 to 1) or two-to-one (2 to 1). The distinctive property of the function lies in the XOR result of two input values that map to the same output, yielding a constant y. If y is all zeros, the function is classified as 1 to 1; otherwise, it is identified as a 2 to 1 function. In classical computing, finding a collision in such a function typically requires $(O(2^{(a-1)} + 1))$ operations, where a is the size of the input. However, Simon's algorithm provides an exponential speedup in comparison. Notably, Simon's algorithm served as an inspiration for Shor's algorithm. This algorithm finds applications, particularly when a suitable query access model is available.

4.4 FUTURE OF CRYPTOGRAPHY: PQC

The recent advancements in quantum computing have triggered significant concerns within the area of cryptography. The advent of quantum algorithms introduces the potential collapse of widely employed techniques for securing digital communications. Consequently, the ongoing responsibility of the cryptographic research community to explore alternative methods has become an urgent imperative. This urgency led to the emergence of PQC, focusing on the examination of PKCs that can withstand the impact of quantum computers. Presently, five families of public key cryptography schemes appear promising in this regard: based on isogeny, hash, code, multivariate, and lattice.

4.4.1 Isogeny-based cryptography

Isogeny-based cryptography schemes are grounded in elliptic curves, defined by the solution of a polynomial equation in two variables. This form of cryptography is a relatively recent development within the broader scope of ECC [12]. Through ongoing research, three significant isogeny-based structures have been proposed for constructing cryptographic schemes: supersingular isogeny Diffie–Hellman (SIDH), ordinary isogeny Diffie–Hellman (OIDH), and commutative SIDH (CSIDH) [13]. The exploration of isogeny-based cryptography began in 2006 [14] when Rostovtsev and Stolbunov introduced the use of isogenies between ordinary elliptic curves for the construction of a cryptography scheme. Their work led to the formulation of the ordinary isogeny problem and the development of OIDH which was designed to create a quantum-resistant PKE and key agreement protocol. In 2011, Jao and coworkers built upon this foundation by proposing the SIDH scheme. This scheme utilized isogenies between supersingular elliptic curves, providing a basis for a cryptosystem applicable to key exchange and encryption. Subsequently, in 2014, Plût and coworkers extended this work to develop a zero-knowledge isogeny identification scheme [15]. Further advancements were made in 2018 when Castryck et al. [16] introduced CSIDH. This structure, based on supersingular elliptic curves, is defined as an endomorphism subring with certain restrictions to supersingular elliptic curves over a prime field, departing from the full endomorphism rings. While isogeny-based cryptography is a relatively young field, the foundational problem of finding isogenies between curves was formulated over a decade ago. Ongoing improvements are essential to bring isogeny-based schemes into practical application.

4.4.2 HB cryptography

HB cryptography relies on one-way hash functions, and any hash function meeting the criteria of a cryptographic hash function can be used in these systems. These cryptographic systems are predominantly employed for creating digital signatures, serving as alternatives to existing signature schemes like RSA. Ralph Merkle introduced HB digital signatures in 1979 [17], and since then, numerous advancements have been made in HB signature schemes. Key properties of HB signature schemes that position them as strong candidates for PQC include minimal security assumptions, flexibility, forward-secure construction, and highly parameterized implementations. There are two main types of HB signatures: stateful and stateless. In stateful schemes, updating the secret key is necessary for signing, and the state of the utilized one-time keys is retained to prevent repetition. Synchronization of the state is crucial for security but entails additional memory overhead. On the other hand, stateless HB signature schemes do not require tracking the state of utilized keys. While they eliminate the state synchronization problem, they exhibit slower performance and higher signature sizes compared to stateful schemes.

HB signatures are categorized into one-time signatures (OTS), few-time signatures (FTS), and many-time signatures (MTS) based on their intended use. OTS [17, 18] cannot be used more than once, FTS [19] allow signing a few novel messages with decreasing security, and MTS permit multiple uses of the same key pair [20]. Many-time schemes often employ a binary tree structure. HB signature schemes have demonstrated robust security against classical and quantum attacks, establishing them as secure and efficient candidates for PQ signature schemes. However, their primary disadvantages include large keys and signature sizes.

4.4.3 Code-based cryptography

Code-based cryptography employs cryptographic algorithms wherein the essential algorithmic component, the foundational one-way function, employs an error-correcting code (EC). McEliece introduced the first code-based PKE scheme in 1978 [21]. In this scheme, the private key is a randomly generated binary undeductible Goppa code, while the public key is determined by randomly permuting the code which results in a randomly generated matrix. The ciphertext is a codeword with added errors, and someone who knows the private key can rectify those errors to determine the original text. Although some parameter adjustments have been made after the introduction of the first code-based cryptography algorithm, no known attacks pose a significant threat to the system, even when considering quantum computers. Several proposals were made to modify McEliece's original scheme to utilize the size of the public key [22–24]. However, many of these modifications were found to be not secure or uneconomical compared to McEliece's original version. Notable among the modifications were the conversions by Kobara and Imai in 2001, which are CCA2-secure and proven to be as secure as the original scheme. These transformations preserve nearly identical transmission rates compared to the original system. The code-based cryptography offers a trade-off between the security of the scheme and the efficiency of the scheme. Various attempts have been made to optimize the original scheme, but finding a balance between these factors remains a challenging task.

4.4.4 Multivariate-based cryptography

Multivariate-based cryptography also known as multivariate PKC (MPKC) [25] involves the examination of PKCs where the trapdoor one-way function adopts the form of a multivariate quadratic polynomial map over a finite field. Typically, the public key *pk* is represented by a set of quadratic polynomials:

$$Pk = \left(pk_1\left(x_1,\ldots,x_n\right),\ldots,pk_m\left(x_1,\ldots,x_n\right)\right)$$

where each pk_i is quadratic nonlinear polynomial in $x = (x_1, \ldots, x_n)$.

The evaluation of these polynomials at any given value corresponds to either the encryption or the verification procedure. Such cryptosystems are referred to as multivariate PKCs. The construction of the first MPKC dates back to 1988 [26]. Notably, the basic construction has remained relatively unchanged for two decades. MPKCs are also occasionally known as trapdoor multivariate quadratic map (MQ) schemes because all current constructions use quadratic public keys for speed, and the inclusion of higher-order terms may offset any potential gain in efficiency. Additionally, higher-order terms might compromise security in the bipolar form. In the current computing landscape, the security of small devices like radio frequency identification (RFID), wireless sensors, personal digital assistants (PDAs), and more have become a growing concern because of the constraints of storage, power, and capacity of communication. Traditional PKCs like RSA are impractical in these settings due to computational complexity. Numerous multivariate signature schemes, like Rainbow [27], present alternatives for small computing devices. Recent research has demonstrated the potential of systems like Rainbow for applications in these devices. Due to their efficiency, the application of MPKCs is a crucial direction, seeking new applications where classical PKCs may not perform satisfactorily. This is likely the domain where MPKCs will exert a discernible influence on real-life applications.

4.4.5 Lattice-based cryptography

A lattice is defined as a set of points in n-dimensional space exhibiting a periodic structure, as illustrated in Figure 4.3. Formally, given n-linearly independent vectors a_1, a_2, ..., $a_i Ri$, the lattice generated by them constitutes the set of vectors.

$$L(a_1, a-2, \ldots, a_i) = (\Sigma i \rightarrow p_n, a_n, p_n \in Z)$$

These vectors a_1, a_2, a_3, ..., a_i are referred to as the lattice's basis.

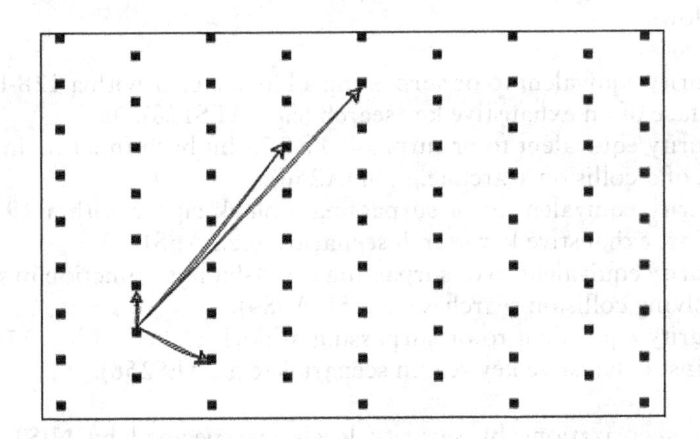

Figure 4.3 A lattice characterized by two dimensions and two potential bases.

The utilization of lattices in cryptography was pioneered in a groundbreaking paper by Ajtai [28]. Cryptographic constructions based on lattices rely on the presumed difficulty of lattice problems, with the shortest vector problem (SVP) being the most fundamental. The LLL algorithm [29], developed by Lenstra et al. in 1982, is the most well-known and extensively studied algorithm for lattice problems. It is a polynomial time algorithm for SVP and other basic lattice problems, achieving the factor of approximation of $2^{o(m)}$, where m denotes the lattice's dimension.

Lattice-based cryptographic constructions can be categorized into two types in terms of security. The first type encompasses practical proposals, often highly efficient but lacking supporting proof of security, such as the GGH scheme [30]. The second type ensures robust provable security grounded in the worst-case complexity of problems of lattice, with only a few, like NTRU [31], demonstrating practical efficiency.

4.5 DEVELOPMENT TOWARD PQC

Following the initiation of the standardization process for PKCs by NIST, substantial advancements have been achieved in the realm of PQC. In 2017, NIST received a total of 82 candidate algorithm submission packages for the PQC standardization process. Out of these, NIST deemed 69 first-round candidates as meeting the requirements for submission and minimum acceptability criteria, qualifying them as "complete and proper submissions," as outlined in FRN-Dec16. This document identified three overarching aspects of assessment criteria that would be employed to assess candidate algorithms throughout the PQC standardization process of NIST. These three aspects encompassed algorithm and implementation features, security, and cost and performance efficiency [32–36]. The assessment of security levels will be organized based on predefined categories established by NIST. These levels, outlined by NIST, are as follows:

1. Security equivalent to or surpassing a block cipher with a 128-bit key in the face of an exhaustive key search (e.g., AES128).
2. Security equivalent to or surpassing a 256-bit hash function in the context of a collision search (e.g., SHA256).
3. Security equivalent to or surpassing a block cipher with a 192-bit key against exhaustive key search scenarios (e.g., AES192).
4. Security equivalent to or surpassing a 384-bit hash function in scenarios involving collision searches (e.g., SHA384).
5. Security equivalent to or surpassing a block cipher with a 256-bit key against exhaustive key search scenarios (e.g., AES256).

These categorizations by security levels, as specified by NIST, provide a structured framework for evaluating the robustness of cryptographic algorithms against various potential threats and attacks. The second criterion involves assessing the cost of the cryptography algorithm. Acknowledging the varied use cases and performance needs, NIST recognizes the potential necessity to standardize multiple algorithms. Within this category, numerous subcategories come into play, such as the size of the keypair, signature, and ciphertext, the computation time of the generation of the keypair, and the failure of decryption. The last criterion involves an examination of the algorithm itself, encompassing its features and implementation. Within this category, three subcategories serve as evaluation criteria:

1. The flexibility of the algorithm
2. Ease of understanding the algorithm
3. Potential of the algorithm across different environments

Recognizing the broader implications of an algorithm's characteristics beyond security and cost, this criterion ensures a comprehensive evaluation that extends to the algorithm's practicality and ease of integration across diverse systems. From the initial pool of 69 first-round candidates, NIST identified and selected 26 candidates to proceed to the second round [37] based on the evaluation criteria and security requirements. Among these 26 candidates, 17 were recognized as key establishment schemes and PKE algorithms, while the remaining 9 were categorized as digital signature schemes, as represented in Table 4.2. This careful selection process reflects NIST's commitment to thorough evaluation and scrutiny in the ongoing PQC standardization

Table 4.2 Schemes selected for the second round

Selected scheme	*Type*
CRYSTALS-Kyber	Lattice-based KEM scheme
FrodoKEM	Lattice-based KEM scheme
LAC	Lattice-based, variant of learning with error (LWE) problem
NewHope	Lattice-based KEM scheme
NTRU	Lattice-based one-way CPA-secure (OW-CPA) PKE scheme
NTRU Prime	Lattice-based deterministic OW-CPA-secure PKE scheme
Round5 (merger of HILA5 and Round2)	Lattice-based KEM and PKE schemes
SABER	Lattice-based KEM and PKE
Three Bears	Lattice-based integer module learning with errors (I-MLWE) KEM
Classic McEliece NTS-KEM BIKE	Code-based KEM

(*Continued*)

Table 4.2 Schemes selected for the second round *(Continued)*

Selected scheme	*Type*
Hamming quasi-cyclic (HQC)	Code-based PKE scheme
LEDAcrypt (merger of LEDAkem and LEDApkc)	Code-based KEM and PKE
Rollo (merger of LAKE, LOCKER, and Ouroboros-R)	Rank-based KEM
RQC	Rank-based PKE algorithm
SIKE	Isogeny-based KEM
CRYSTALS-Dilithium	Lattice-based digital signature scheme
FALCON	
qTESLA	
GeMSS	Multivariate-based digital signature scheme
LUOV	
MQDSS	
Rainbow	
Picnic	Non-interactive zero-knowledge digital signature scheme
SPHINCS+	HB digital signature scheme

process. Moving forward to the standardization process's third round, NIST opted for 15 second-round candidates [38]. Out of these 15 selected candidates, 7 were chosen as finalists, while the remaining 8 secured positions as alternative candidates. The list of selected and alternative candidates is given in Table 4.3. The group of finalists is anticipated to be prepared for standardization shortly after the conclusion of the third round. The alternative candidates were viewed as potential options for future standardization, possibly following another round of evaluation. After the third round [39], NIST chose CRYSTALS-Kyber as the key encapsulation mechanism (KEM) for standardization [40–42]. The digital signature schemes selected for standardization include CRYSTALS-Dilithium, Falcon, and SPHINCS+.

Table 4.3 Finalist schemes selected for the third round

Finalist	
PKE/KEM algorithm	Security level
Classic McEliece	5
CRYSTALS-Kyber	1,3,5
NTRU	1
SABER	1,3,5
Digital signature scheme	**Security level**
CRYSTALS-Dilithium	1,2,3
FALCON	1,5
Rainbow	1,3,5

Table 4.4 Alternative schemes selected for third round

Alternative	
PKE/KEM algorithm	Digital signature scheme
BIKE	GeMSS
FrodoKEM	Picnic
HQC	SPHINCS+
NTRU Prime	
SIKE	

Four additional KEMs, namely, BIKE, Classic McEliece, HQC, and supersingular isogeny key encapsulation (SIKE), will continue to undergo evaluation, with NIST anticipating the standardization of at least one of them after the fourth round. Table 4.3 displays all the finalist schemes chosen after the third round, while Table 4.4 showcases alternative schemes.

4.5.1 Development of quantum computers

Quantum computers embody a revolutionary approach to computing, leveraging the principles of quantum mechanics to execute specific types of calculations significantly more efficiently than classical computers [43]. IBM made significant strides in the field of quantum computing (Table 4.5). In 2016, they placed the first quantum computer on the cloud, and the following year, they

Table 4.5 Roadmap for the development of quantum computers as per IBM [45]

Year	*Objective*	*Processor used*	*Number of qubits*	*Key advancement*
2019	Execute quantum circuits using the IBM cloud	Falcon	27	Optimized lattice
2020	Illustrate and create prototypes for quantum algorithms and applications	Hummingbird	65	Extensible readout
2021	Accelerate the execution of quantum programs by a factor of 100 within the Qiskit environment	Eagle	127	Innovative packaging and control mechanisms
2022	Integrate dynamic circuits into Qiskit runtime to unlock additional computational capabilities	Osprey	433	Reduced size of the components
2023	Optimizing applications through the implementation of elastic computing and parallelization within Qiskit runtime	Condor	1,121	Integration

(Continued)

Table 4.5 Roadmap for the development of quantum computers as per IBM [45] *(Continued)*

Year	*Objective*	*Processor used*	*Number of qubits*	*Key advancement*
2024	Enhance the precision of Qiskit runtime through the implementation of scalable error mitigation techniques	Flamingo	1,386+	Build new infrastructure and quantum error correction
2025	Expand the scalability of quantum applications using the Circuit Knitting toolbox to manage Qiskit runtime	Kookaburra	4,158+	
2026	Elevate the accuracy and speed of quantum workflows by integrating error correction within the Qiskit environment		Scaling 10–100k qubits	

introduced an open-source software development kit, Qiskit, designed for programming these quantum computers. The unveiling of the IBM Quantum System One in 2019 marked the introduction of the first integrated quantum computer system. Subsequently, in 2020, IBM released a development roadmap outlining their strategy to advance quantum computers into a commercial technology. In 2021, IBM achieved a milestone with the introduction of the 127-qubit IBM Quantum Eagle processor. They also launched Qiskit Runtime, a runtime environment that integrates classical and quantum systems, supporting the containerized execution of quantum circuits with enhanced speed and scalability [44]. Building on the design principles established for smaller processors, IBM continued its progress by unveiling the 433-qubit IBM Quantum Osprey processor in 2022. In 2023, IBM plans to unveil the 1,121-qubit IBM quantum based on the Condor processor. This processor will integrate insights gained from previous models, aiming to further reduce critical 2-qubit errors. The objective is to enable the execution of longer quantum circuits, enhancing the capabilities of quantum computing technology.

4.6 FUTURE PROSPECTS AND CHALLENGES

PQC is an emerging research topic expected to play a pivotal role in the years to come. Within this field, there are five major families upon which cryptosystems are being constructed. Each of these families presents its own set of advantages and disadvantages, encompassing factors such as bit security, size, memory overhead, and computational requirements. Researchers in the field of PQC are actively seeking new mathematical problems that pose challenges for quantum computers. These challenges serve as foundations for developing novel cryptosystems. Simultaneously, researchers are exploring problems requiring minimal overhead and demanding fewer computational resources. This dual approach aims to fortify cryptographic systems against the potential threats of quantum computers, ensuring robust security in the PQ era. One additional challenge lies in addressing cryptographic flexibility in legacy devices that may be impractical to reconstruct. It is crucial to undertake research and gain a comprehensive understanding of the individuals involved in the process, as well as the existing policies. An essential aspect to consider is the establishment of the right incentives for vendors and developers of software products to integrate cryptographic flexibility into all necessary products. Customers and governments play pivotal roles in shaping this aspect, as customers hold the power to exert pressure on vendors, being the primary source of business for them. Creating a supportive environment for cryptographic agility requires collaboration between various stakeholders, with a focus on aligning incentives and ensuring the adaptability of cryptographic solutions across diverse systems and devices.

Creating algorithms for quantum computers necessitates a distinct approach compared to classical computing. The development of a robust quantum software ecosystem, encompassing programming languages and tools, is imperative for facilitating widespread adoption and efficient utilization of quantum computing resources. As quantum computing technology advances, the establishment of standardized and user-friendly tools becomes increasingly crucial to empower researchers and developers to harness the full potential of quantum systems. This includes the design and implementation of quantum algorithms, optimization techniques, and programming frameworks that cater to the unique principles and capabilities of quantum computing. Building a well-defined quantum software ecosystem is key to unlocking the transformative power of quantum computing across various domains and industries. Researchers can play an active role in the advancement of lightweight PQC, particularly in response to the growing prevalence of IoT (Internet of Things) devices. The demand for lightweight cryptography has become imperative in this context. Lightweight PQC constitutes a crucial field that warrants dedicated research efforts. The focus is on creating algorithms that are not only quantum-safe but also impose minimal overhead. As the landscape of technology continues to evolve, the emphasis on lightweight cryptographic solutions becomes increasingly significant, aligning with the specific constraints and requirements of IoT devices and other resource-constrained environments. Quantum cryptanalysis and the development of new quantum algorithms represent a relatively underexplored research area. Shor's groundbreaking work presented two quantum algorithms capable of breaking the security of current asymmetric algorithms used in present-day cryptography. Similarly, PQC algorithms may face potential threats from different quantum algorithms, which could compromise their security. Therefore, there is a pressing need for more research in this field to understand and mitigate the possible risks posed by quantum computing to cryptographic systems. Investigating quantum cryptanalysis and exploring new quantum algorithms will be essential to developing robust cryptographic solutions that can withstand the capabilities of quantum computers. This research will be instrumental in ensuring the security and resilience of cryptographic systems in the stage of quantum computing.

4.7 CONCLUSION

The exploration of quantum computing has unraveled a paradigm shift in the security landscape of multimedia through cryptography. Classical cryptographic schemes, once considered robust, are now susceptible to quantum algorithms, such as Shor's, which pose unprecedented threats to the confidentiality and integrity of multimedia data and this emphasizes the urgency to transition toward quantum-resistant solutions which leads us toward PQC. The introduction of PQC emerges as a pivotal shift, offering a promising avenue for the future of secure communication. The exploration of diverse cryptographic families, including isogeny-based, multivariate-based, code-based, HB, and lattice-based approaches, reflects the ongoing efforts to develop robust alternatives that can be used for quantum computing. Our detailed exploration of the NIST standardization process of cryptography schemes sheds light on the rigorous evaluation and selection of public key cryptography schemes and digital signature schemes. The distinction between finalists and alternatives highlights the meticulous consideration of security, efficiency, and standardization requirements. The timeline of quantum computer development, outlined in the provided table, highlights the swift advancements in quantum technology. This evolution necessitates a proactive stance in fortifying cryptographic systems against potential quantum threats. As we stand on the brink of this quantum revolution, it is crucial to acknowledge the challenges inherent in transitioning from classical to quantum computing. Addressing these challenges, such as cryptographic agility and the need for lightweight solutions, is pivotal for a seamless integration into the quantum era. Looking ahead, the future of cryptography lies in the hands of researchers who can navigate the complexities of quantum computing. Opportunities abound in developing novel quantum-resistant algorithms, refining PQ cryptographic solutions, and unraveling the mysteries of quantum-safe communication. The development of cryptography schemes adaptable to both classical and quantum computers heralds an exciting frontier for cryptographic research and innovation.

REFERENCES

1. Multimedia Security Using Encryption: A Survey, Hosny, Khalid M, Zaki, Mohamed A, Lashin, Nabil A, Fouda, Mostafa M, and Hamza, Hanaa M, IEEE Access, IEEE, 2023

2. Post quantum cryptography: Techniques, challenges, standardization, and directions for future research, Bavdekar, Ritik, Chopde, Eashan Jayant, Bhatia, Ashutosh, Tiwari, Kamlesh, Daniel, Sandeep Joshua et al., arXiv:2202.02826, 2022
3. Status Report on the First Round of the NIST Post-Quantum Cryptography Standardization Process, Alagic, Gorjan, Alagic, Gorjan, Alperin-Sheriff, Jacob, Apon, Daniel, Cooper, David, Dang, Quynh, Liu, Yi-Kai, Miller, Carl, Moody, Dustin, Peralta, Rene et al., US Department of Commerce, National Institute of Standards and Technology, 2019
4. Post quantum cryptography: A review of techniques, challenges and standardizations, Bavdekar, Ritik, Chopde, Eashan Jayant, Agrawal, Ankit, Bhatia, Ashutosh, and Tiwari, Kamlesh, 2023 International Conference on Information Networking (ICOIN), 146–151, IEEE, 2023
5. Quantum computing: The risk to existing encryption methods, Kirsch, Zach and Chow, Ming. Retrieved from: http://www.cs.tufts.edu/comp/116/archive/fall2015/zkirsch.pdf, 2015
6. The discrete logarithm problem, McCurley, Kevin S, Proceedings of Symposia in Applied Mathematics, 42, 49–74, 1990
7. Consumer applications of quantum computing: A promising approach for secure computation, trusted data storage, and efficient applications, Humble, Travis, IEEE Consumer Electronics Magazine, 7, 6, 8–14, 2018
8. The impact of quantum computing on present cryptography, Mavroeidis, Vasileios, Vishi, Kamer, Zych, Mateusz D, and Jøsang, Audun, arXiv:1804.00200, 2018
9. An overview of quantum cryptography and Shor's algorithm, Ugwuishiwu, CH, Orji, UE, Ugwu, CI, and Asogwa, CN, International Journal of Advanced Trends in Computer Science and Engineering, 9, 5, 2020
10. A fast quantum mechanical algorithm for database search, Grover, Lov K, Proceedings of the Twenty-Eighth Annual ACM Symposium on Theory of Computing, 212–219, 1996
11. Simon's Algorithm, Baaquie, Belal Ehsan and Kwek, Leong-Chuan, Quantum Computers: Theory and Algorithms, 203–207, Springer, 2023
12. The elliptic curve Diffie–Hellman (ECDH), Haakegaard, Rakel and Lang, Joanna. Available at: https://koclab.cs.ucsb.edu/teaching/ecc/project/2015Projects/Haakegaard+Lang.pdf, 2015
13. Isogeny-based cryptography: A promising PQ technique, Peng, Cong, Chen, Jianhua, Zeadally, Sherali, and He, Debiao, IT Professional, 21, 6, 27–32, 2019
14. Public-Key Cryptosystem Based on Isogenies, Rostovtsev, Alexander and Stolbunov, Anton, Cryptology ePrint Archive, 2006
15. Towards quantum-resistant cryptosystems from supersingular elliptic curve isogenies, De Feo, Luca, Jao, David, and Plût, Jérôme, Journal of Mathematical Cryptology, 8.3, 209–247, 2014
16. CSIDH: An efficient PQ commutative group action, Castryck, Wouter, Lange, Tanja, Martindale, Chloe, Panny, Lorenz, and Renes, Joost, Advances in Cryptology–ASIACRYPT 2018: 24th International Conference on the Theory and Application of Cryptology and Information Security, Brisbane, QLD, Australia, December 2–6, 2018, Proceedings, Part III, 24, 395–427, Springer, 2018
17. Secrecy, Authentication, and Public Key Systems., Merkle, Ralph Charles, Stanford University, 1979
18. Tuning the Winternitz hash-based digital signature scheme, Perin, Lucas Pandolfo, Zambonin, Gustavo, Martins, Douglas Marcelino Beppler, Custodio, Ricardo, and Martina, Jean Everson, 2018 IEEE Symposium on Computers and Communications (ISCC), 00537–00542, IEEE, 2018

19. Better than BiBa: Short one-time signatures with fast signing and verifying, Reyzin, Leonid and Reyzin, Natan, Australasian Conference on Information Security and Privacy, 144–153, Springer, 2002
20. SPHINCS, Aumasson, Jean-Philippe, Bernstein, Daniel J, Beullens, Ward, Dobraunig, Christoph, Eichlseder, Maria, Fluhrer, Scott, Gazdag, Stefan-Lukas, Hülsing, Andreas, Kampanakis, Panos, Kölbl, Stefan, et al., 2019
21. A public-key cryptosystem based on algebraic coding theory, McEliece, Robert J, DSN Progress Report, 42–44, 114–116, 1978
22. Reducible rank codes and their applications to cryptography, Gabidulin, Ernst M, et al., IEEE Transactions on Information Theory 49.12, 3289–3293, 2003.
23. Ideals over a non-commutative ring and their application in cryptology, Gabidulin, Ernst M, Paramonov AV, and Tretjakov, OV, Advances in Cryptology—EUROCRYPT'91: Workshop on the Theory and Application of Cryptographic Techniques Brighton, UK, April 8–11, 1991 Proceedings 10, Springer, Berlin, 1991.
24. Shorter keys for code based cryptography, Gaborit, Philippe, Proceedings of the 2005 International Workshop on Coding and Cryptography (WCC 2005), 81–91, 2005
25. Progress in multivariate cryptography: Systematic review, challenges, and research directions, Dey, Jayashree and Dutta, Ratna, ACM Computing Surveys, 55, 12, 1–34, 2023
26. Public quadratic polynomial-tuples for efficient signature-verification and message-encryption, Matsumoto, Tsutomu, and Imai, Hideki, Advances in Cryptology—EUROCRYPT'88: Workshop on the Theory and Application of Cryptographic Techniques Davos, Switzerland, May 25–27, 1988 Proceedings, 7, 419–453, Springer, 1988
27. Rainbow, a new multivariable polynomial signature scheme, Ding, Jintai and Schmidt, Dieter, International Conference on Applied Cryptography and Network Security, 164–175, Springer, 2005
28. Generating hard instances of lattice problems, Ajtai, Miklós, Proceedings of the Twenty-Eighth Annual ACM Symposium on Theory of Computing, 99–108, 1996
29. Factoring polynomials with rational coefficients, Lenstra, Arjen K, Lenstra, Hendrik Willem, and Lovász, Ĺaszĺo, Mathematische annalen, 261, 515–534, 1982
30. Public-key cryptosystems from lattice reduction problems, Goldreich, Oded, Goldwasser, Shafi, and Halevi, Shai, Advances in Cryptology—CRYPTO'97: 17th Annual International Cryptology Conference Santa Barbara, California, USA August 17–21, 1997 Proceedings 17, 112–131, Springer, 1997
31. NTRU: A ring-based PKC, Hoffstein, Jeffrey, Pipher, Jill, and Silverman, Joseph H, International Algorithmic Number Theory Symposium, 267–288, Springer, 1998
32. On the Development and Standardisation of Post-Quantum Cryptography: A Synopsis of the NIST Post-Quantum Cryptography Standardisation Process, Its Incentives, and Submissions, Legernaes, Maja Worren, NTNU, 2018
33. Multimodal imputation-based stacked ensemble for prediction and classification of air quality index in Indian cities, Rao, RS, Kalabarige, LR, Alankar, B, and Sahu, AK, Computers and Electrical Engineering, 114, 109098, 2024
34. Improved multiview biometric object detection for anti spoofing frauds, Asmitha, P, Rupa, C, Nikitha, S, Hemalatha, J. and Sahu, AK, Multimedia Tools and Applications, 1238, 1–17, 2024. https://doi.org/10.1007/s11042-024-18458-8
35. A novel and secure fake-modulus based Rabin-3 cryptosystem, Ramesh, R.K., Dodmane, R., Shetty, S., Aithal, G., Sahu, M. and Sahu, A.K., Cryptography, 7.3, 44, 2023

36. Chaotic-map based encryption for 3D point and 3D mesh fog data in edge computing, Raghunandan, K.R., Dodmane, R., Bhavya, K., Rao, N.K. and Sahu, A.K., IEEE Access, 11, 3545–3554, 2022
37. Status Report on the First Round of the NIST Post-Quantum Cryptography Standardization Process, Alagic, Gorjan, Alagic, Gorjan, Alperin-Sheriff, Jacob, Apon, Daniel, Cooper, David, Dang, Quynh, Liu, Yi-Kai, Miller, Carl, Moody, Dustin, Peralta, Rene, et al., 2019, US Department of Commerce, National Institute of Standards and Technology
38. Status Report on the Second Round of the NIST Post-Quantum Cryptography Standardization Process, Alagic, Gorjan, Alperin-Sheriff, Jacob, Apon, Daniel, Cooper, David, Dang, Quynh, Kelsey, John, Liu, Yi-Kai, Miller, Carl, Moody, Dustin, Peralta, Rene, et al., US Department of Commerce, NIST, 2, 2020
39. Status Report on the Third Round of the NIST Post-Quantum Cryptography Standardization Process, Alagic, Gorjan, Apon, Daniel, Cooper, David, Dang, Quynh, Dang, Thinh, Kelsey, John, Lichtinger, Jacob, Miller, Carl, Moody, Dustin, Peralta, Rene et al., US Department of Commerce, NIST, 2022
40. Digital image steganography techniques in spatial domain: A study, Sahu, A.K. and Sahu, M., International Journal of Pharmacy & Technology, 8.4, 5205–5217, 2016
41. A novel hybrid authentication protocol utilizing lattice-based cryptography for IoT devices in fog networks, Roy, KS, Deb, S, and Kalita, HK, Digital Communications and Networks, 2022. https://doi.org/10.1145/3614433
42. NMRMG: Nonlinear multiple-recursive matrix generator design approaches and its randomness analysis, Deb, S, Pal, S, and Bhuyan, B, Wireless Personal Communications, 125.1, 577–597, 2022
43. Quantum computing and cryptography, Easttom, Chuck, Modern Cryptography: Applied Mathematics for Encryption and Information Security, 397–407, Springer, 2022
44. IBM's Qiskit tool chain: Working with and developing for real quantum computers, Wille, Robert, Van Meter, Rod, and Naveh, Yehuda, 2019 Design, Automation & Test in Europe Conference & Exhibition (DATE), 1234–1240, IEEE, 2019
45. IBM's Roadmap for Scaling Quantum Technology, Gambetta, Jay, IBM Research Blog, 2020

Chapter 5

Secured textual medical information using a modified LSB image steganography technique

Roseline Oluwaseun Ogundokun, Oluwakemi Christiana Abikoye, Ezekiel Adebayo Ogundepo, Akinbowale Nathaniel Babatunde, Abdul Rahman Tosho Abdulahi, and Aditya Kumar Sahu

5.1 INTRODUCTION

An electronic health record (EHR) is an organized group of automated medical reports concerning a convalescent person or people [1]. It might also comprise medical checkups, patient demographics, detection observations, treatments, medicines, previous health records, histological details, and then further discoveries, fundamental symptoms, vaccinations, laboratory information, and radiology details [2]. One of the unified grounds considered in information hiding and information security is the interchange of information, data, or knowledge through protected or hidden media (cover media). Numerous discrete methods, such as cryptography and encryption, have been developed to encrypt and decrypt information sequentially to retain a proportion of the communication secret. Despite that, steganography has some upper hands more than these methods; it keeps the being of the communicated mystery together with the confidential information.

Delicate data transmitted through a network must be protected safely from inspection or intrusion attempts. In contrast, an individual's identification must be confirmed [3] and validated before admission to the safe data/information is granted [4,5]. The principal motive of steganography in information security is to safeguard information or data from completely unlicensed entry, use, alteration, or erasing. Countless surveys have centered on keeping secrets of data, information, or messages to guarantee and not vulnerable the data to any danger that may affect them. The field of biomedical sciences has heavily relied on the data elicited from their patients for drug prescription and treatment. These data are the confidential property of the patient, must be used for their intended purpose, and must not be revealed to unauthorized non-medical personnel. However, due to the need for medical consultation, occasions may warrant transferring elicited medical information within or outside a medical facility. Furthermore, medical information may be transmitted over a network for educational and research purposes [6]. This information sharing among

 DOI: 10.1201/9781032663647-5

medical personnel and possible transmission over a network has opened medical information to attack threats. This threat can be minimized by securing medical information before dispatch or securing the channel of communication. Cryptography and steganography are the most adopted methods for securing medical information before transmission [7].

Nevertheless, before adopting any stenographic technique, the secret image must first be encrypted [8]. Since a picture is known to be composed of pixels that connote the intensities of light at different points, least significant bit (LSB) steganography works by swapping the LSBs of the concealment object pixels with that of the undisclosed information such that the changes are not perceptible by or visible to the individual vision [9]. Though the LSB technique is the utmost steganography method, literature has reported that it is susceptible to statistical attacks [9–11]. Privacy protection is necessary for health industries [12,13]. Usually, therapeutic data should comprise images and patient records [12]. This information ought to be reserved privately to uphold the confidentiality of patients [14,15]. Due to weaknesses in digital data properties, traditional LSB schemes over textual and image steganography cannot be extended, especially to health industry datasets. Therefore, a protected keyframe abstraction technique is obligatory to guarantee appropriate, precise confidentiality and conservation of medical facilities. In addition, obtaining a satisfactory degree of protection in a lucrative manner is fundamentally problematic when considering the shortcomings of health systems in real time.

Different steganography techniques in the spatial domain have been presented, such as the LSB substitute [16], optimal pixel adjustment process (OPAP) [17], the pixel value difference (PVD) [18], and modulus function–pixel value difference (MF-PVD) [19]. These techniques have low visual quality and limited capacity [20]. Since the steganography technique provides security by concealing the undisclosed communication in a cover object, the mission fails when the adversary detects the secret message. The standard LSB steganography technique's weakness is that it possesses a big payload and is simple to see and beat. They are also not resilient against lossy compression and image filters; image resaving loses personal data but always compensates for the image's statistical properties [21].

Though the LSB approach takes advantage of the human vision sensitivity (HVS) vulnerability to mask confidential data in a concealment object in a mode that the person's eyes can't detect, the arithmetical features of the resulting stego objects show abundant alteration of the initial concealment object that undermines the protection of the concealed record. Since it utilizes solitary 3 bits in a distinct pixel or a single bit for each color object, its hiding ability is also comparatively limited. According to Provos and Honeyman [22], a steganography treatment fails if it causes anyone to mistrust the carrier medium. With the massive volume of daily data gathered, accessed, and shared electronically worldwide, there is a strong need to improve data protection algorithms regularly. Hence, steganographic data-hiding algorithms and methods must improve their steganalysis imperceptibility and data-hiding ability.

This chapter suggests a modified LSB approach that uses several shifts for implanting medical information in a concealment object. This modified approach assisted in boosting and intensifying both the imperceptibility and the concealing potential of the standard LSB scheme. This is to aid in preventing the likelihood of the positions where the confidential message is covered in the concealment object, consequently increasing the imperceptibility by lowering the difference between the earliest concealment object's statistical characteristics and those of the stego object. To handle extra data and thus increase the hiding power, the sum of bits employed for each color medium could also be diverse. This chapter aims to improve the imperceptibility and concealing ability of the traditional LSB steganography entrenching system by modifying how the medical information is implanted in the concealment object's LSB swapped during the embedding process with the significant bits of the hidden data.

Therefore, this chapter proposes to modify the existing LSB steganography technique by employing logical bit shift operations. This works by shifting the LSB of the red (R), green (G), and blue (B) components of the concealed object pixels a predetermined number of times. The bits from the unrevealed communication would then replace the changed bits. As a result, this chapter intends to evaluate earlier investigations on digital textual steganography using LSB steganography as an information-concealing method, to examine the perceptibility measurement in the information structure of the standard LSB procedure, to inspect the effect of imperceptibility and hiding potentials by using the diverse and arbitrarily selected bits of the concealment object in the LSB steganography system entrenching procedure, and to modify the LSB standard algorithm to perform better for hiding capacity.

5.1.1 Research questions

This study explores and evaluates various aspects of steganography, explicitly focusing on the LSB technique. The research questions guiding this investigation are designed to understand the current landscape, assess performance metrics, and examine the effects and improvements of the LSB method. Here are the research questions and the rationale behind each:

1. What is the present deduction about steganography and LSB techniques in the literature?

   This question aims to gather and summarize the current understanding and findings in steganography, particularly the LSB technique. By reviewing existing literature, we can identify trends, gaps, and the overall state of knowledge in this area.
2. What are the state-of-the-art performance evaluation values?

   Evaluating the performance of current state-of-the-art steganography techniques is crucial for benchmarking and understanding the effectiveness of different methods. This question seeks to compile and analyze these performance metrics for a comparative baseline.

3. What is the effect of LSB steganography on stego image?

   This question addresses the impact of applying LSB steganography on the quality and integrity of the stego image. Understanding these effects is essential for assessing hidden information's practicality and visual perceptibility.

4. What is the effect of the modified LSB steganography proposed?

   Here, we investigate the outcomes of implementing a modified version of the LSB technique. The goal is to determine whether the proposed modifications improve the performance or mitigate any identified drawbacks of the standard LSB method.

5. What are the modified LSB peak signal-to-noise ratio (PSNR) and mean squared error (MSE) values?

   PSNR and MSE are standard metrics used to measure image quality. This question evaluates these values for the modified LSB technique to quantify its effectiveness and compare it to existing methods.

The remaining chapter is prepared thus: Section 5.2 discusses the prevailing LSB steganography techniques and the related works. In Section 5.3, the author postulated a modified LSB image steganography. In Section 5.4, the authors discuss the findings of implementing the proposed system. Section 5.5 is about the performance examination of the postulated method, and Section 5.6 discusses the discovery from the results and the evaluation. Section 5.7 concludes the chapter.

5.1.2 Existing LSB techniques

The steganography methods can be classified into seven categories, even though, in some cases, accurate or precise classification is impossible, as shown in Figure 5.1.

Habeeb [23] postulated a novel approach to safeguard communication from hacking. They used the method of hiding, which is steganography.

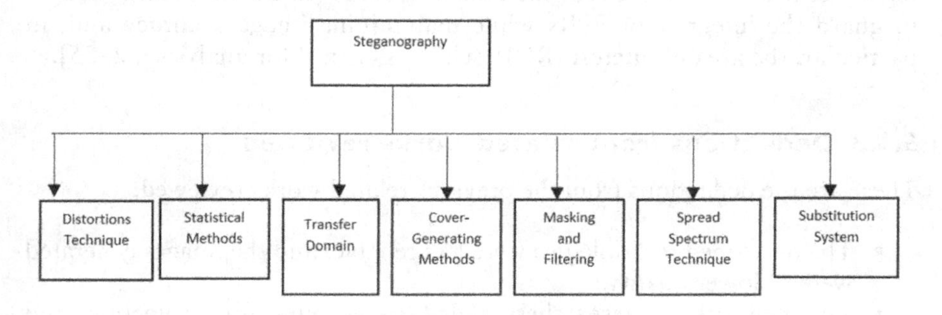

Figure 5.1 Different methods of steganography.

They used the encryption method to encode the text by employing the password amid the source and the destination, after which they used the LSB approach to hide the text in the image (cover object). MATLAB was used to implement the system. Masoud and Ghazi [24] postulated an improved least significant digit (LSD) digital watermark technique. Optimizing choice reduces the number of pixels whose value will be changed, and their system's drawback is that exact efficiency cannot be achieved by haphazard pixel value selection. Ali and Doegar [25] suggested LSB-based steganography using a parity checker. The advantage of their system is its embedding capacity and the system's weakness is that there exists less robust and lesser quality image. Juneja and Sandhu [26] proposed a hybrid feature recognition method. Two constituents built replacement procedures and adaptive LSB. The system improved imperceptibility and better resistance to various steganalysis attacks, and the system limitation was that capacity and robustness were not considered in their research.

Juneja and Sandhu [27] projected two components of LSB procedures for enclosing hidden documents in the LSBs of blue parts and partial green features. Their method was a flexible LSB-centered steganography for inserting documents established on records and advanced encryption standard (AES) in the proposed method, making it more robust. Performance evaluation of the research work was not conducted in this research. Gupta et al. [28] examined a new LSB method, the enhanced least significant bit (ELSB) because it was discovered that the existing LSB algorithm has been inspected and found to have an extra volume of alteration. Their scheme reduces the alteration level of the image, which is inattentive to a person's eye, but possesses a limitation in that the security and imperceptibility level of the image was not looked into. Chawla et al. [29] suggested an Innovative procedure for image steganography in the spatial domain employing the last 2 bits of the pixel value. The study gave a limited alteration in concealment object, but the limitation of the study is that there was distortion in the cover image quality. Sahu and Swain [30] proposed a survey to increase entrenching competence by employing a double-layer entrenching technique and reducing the falsification created by the stego image to enhance its accuracy. Al-Dmour and Al-Ani [31] wanted to include the confidential specifics of their patients' diagnoses in their photographs. It hides coded electronic patient records (EPRs) in medical images to guard the integrity of EPRs while maintaining image accuracy and, in particular, the area of interest (ROI), which is critical for analysis [32–35].

5.1.3 Deductions from related works reviewed

These are the deductions from the previous related works reviewed:

- The developed technologies were less reliable, and the images generated were of lower quality.
- The majority of researchers failed to account for protection and misperceptions.

- Only a few researchers took capability and robustness into account.
- In some previous studies, distinguishing between concealment and stego picture was difficult.
- Some studies failed to account for the state of higher PSNR and lower MSE.
- The cover image used in some experiments was distorted, resulting in poor stego image quality, making it easier for intruders to spot a protected message being transmitted.
- In some prior studies, distortion, robustness, and imperceptibility were not considered.

With all these conclusions, the research altered the previous typical LSB steganography procedures to address the issues of imperceptibility, less robustness, falsification, and safety.

5.2 MATERIALS AND METHOD

This section discusses the material and method used. This section also deals with the postulated system algorithm and the framework.

5.2.1 Proposed system framework

The framework to be designed for the secured medical information system using the modified LSB steganography technique is shown in Figure 5.2. The framework can be divided into the encryption stage and the decryption stage. Each stage can also be subdivided into four components:

i. The secret message
ii. The circular shift LSB algorithm
iii. The cover medium
iv. The stego image

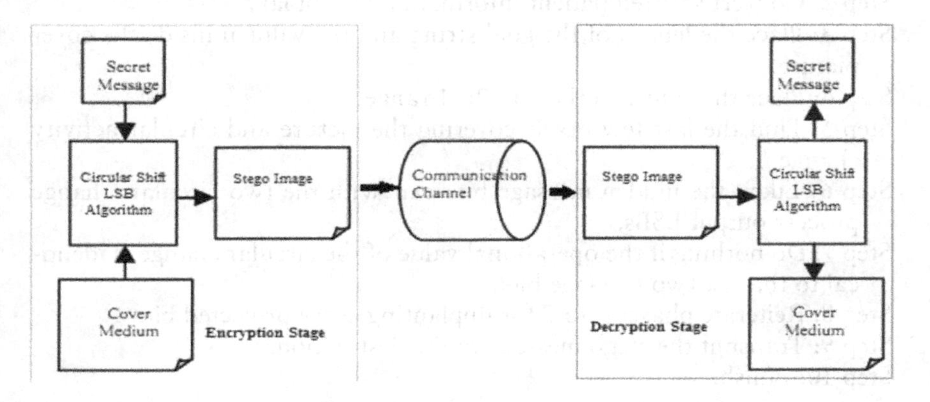

Figure 5.2 Framework for the secured medical information system.

5.2.2 Proposed system

This chapter projected an LSB entrenching technique that employs diverse and randomly selected bits from the concealment object to mask the undisclosed message. This chapter focuses on using the steganography of digital images in data-hiding systems. The approach presented in this analysis is not generally appropriate for other cover newspapers since it is primarily developed for digital photographs. Improving the perceptibility and data-hiding capabilities of the LSB steganography data hiding system is the primary concern of the proposed steganographic applications. New digital communication's increasingly rising possibilities desperately need more reliable means of securing information counter to illegal entrée during transmission. The augmented necessity to protect digital communication and information ensures that in the years to come, studies in this field will begin to draw more interest. Steganography, a method for inconspicuously hiding hidden data within a host dataset, provides improved security possibilities in open system environments [16]. This chapter aims to contribute to the field of digital image steganography and, precisely, to the perceptibility of the LSB procedure counter to steganalysis and its data-hiding ability by suggesting the utilization of an entrenching mechanism that uses numerous and pseudorandomly chosen concealment object bits to mask the hidden details [21].

5.2.3 Proposed system algorithms

5.2.3.1 Process for entrenching the undisclosed message using circular shift LSB

The following steps will be taken to create the stego image:

Step 1: Input textual patient information as the secured message.
Step 2: Convert secured patient information into binary.
Step 3: Place the length of the goal string and the value n inside the cover picture.
Step 4: Place the values in the 0 to 2n-1 range
Step 5: Find the last few pixels covering the picture and circular activity change.
Step 6: Liken the hidden message bits pair with the two circular Change process output LSBs.
Step 7: Do nothing if the operational value of the circular change is identical to the last two message bits.
Step 8: Reiterate phases 4 to 7 for implanting every projected bit.
Step 9: Transmit the stego message to the destination.
Step 10: Finish.

5.2.4 Algorithm for retrieving the undisclosed message from the stego object

Step 1: Accept the stego message as input.
Step 2: Find the last n bits of the stego picture that the sender collects from every target pixel of the stego-image and conduct circular shift operations.
Step 3: Twofold LSBs of circular shift activity are the hidden data bits from the end.
Step 4: Concatenate. Seven bits and translate them to their respective decimal value.
Step 5: Store it as a character, type it into another file and enter the goal message.
Step 6: End.

The steganography framework designed for this study will be implemented using the MATLAB 2018a programming environment. MATLAB is chosen for its robust capabilities in handling image processing tasks, ease of use for matrix computations, and extensive libraries that support the development of steganographic algorithms.

5.2.4.1 Technical specifications of the implementation system

To ensure smooth and efficient implementation of the framework, the following hardware specifications are utilized:

Processor: Intel Core i7

- The Intel Core i7 processor is selected for its high performance and ability to efficiently handle complex computations and large datasets. Its multiple cores and high clock speed ensure the implementation can process image data and perform necessary calculations swiftly.

Memory (RAM): 8GB

- An 8GB RAM provides ample memory for running MATLAB 2018a and handling large images and datasets. This memory capacity helps minimize lag and ensure the system can multitask effectively without running into memory bottlenecks.

Storage Capacity: 500 GB

- A 500 GB hard drive offers sufficient storage space for the software, the operating system, and the large image files used in the implementation. It ensures that there is adequate room to store various versions of the stego images and other related data without running out of space.

5.3 FINDINGS AND DISCUSSION

The authors employed MATLAB to implement the projected procedures where two-dimensional concealment objects were employed to insert patients' medical data. The structure also used text with different sizes. When the algorithms were implemented, it was discovered that it was hard to differentiate between the initial concealment object and the stego message (the thing hiding the texts). The results from the implementation are discussed thus.

5.3.1 The method entrenching stage

Like most steganography schemes, the modified LSB steganography solution uses a GUI to make the users' experience as simple as possible. Figure 5.3 is a representation of the main window interface. The control panel of this built framework is the main GUI window through which the user may perform different tasks. The handler initially inputs the secured medical textual information. The user then types in the stego key (secured key) and picks the concealment object for implanting the medical message, as shown in Figure 5.5. The application models also permit glancing for the concealment object in the computer storage or peripheral gadgets. Then, the system produced the stego image displayed on the application interface and saved it in a document on the computer. The output interface now shows both the original concealment image and the stego image (image containing the implanted medical information).

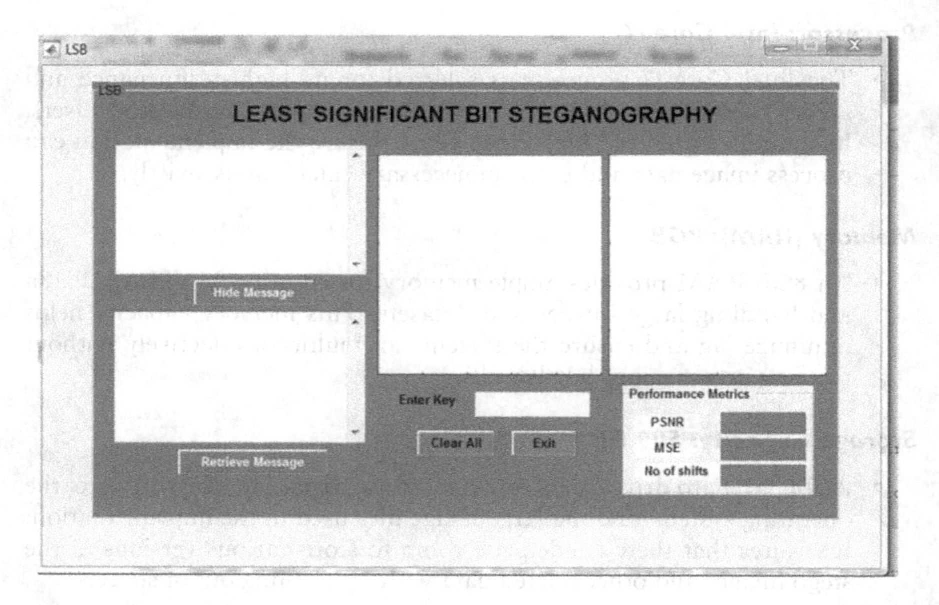

Figure 5.3 Interface of the method.

The suggested method interface for a stable medical information system is seen in Figure 5.3. The figure shows the graphical user interface (GUI) for a LSB steganography application designed to hide and retrieve messages within images. The interface includes text boxes for inputting and displaying messages, buttons for hiding and retrieving messages, areas for showing the original and stego images, and fields for entering a security key. Additionally, it features control buttons to clear fields or exit the application, as well as performance metrics such as PSNR and MSE to assess the quality of the steganography process. The functional flow involves embedding a message into an image and extracting it, with metrics provided for quantitative feedback.

Figure 5.4 illustrates the method interface, which displays the patient's medical history in text format and the stego key (which is only understood by the sender and receiver) being inputted. The figure shows a GUI for a LSB steganography application designed to hide and retrieve sensitive information within images. The GUI is labeled "LEAST SIGNIFICANT BIT STEGANOGRAPHY" and includes a text box on the left with a sample patient's record containing age, zip code, income, and disease information, indicating the hidden data type. Below this are the "Hide Message" and "Retrieve Message" buttons for embedding and extracting hidden information. The center of the GUI features two large blank areas for displaying the original cover image and the stego image. An "Enter Key" field is provided for additional security during steganography. Control buttons labeled "Clear All" and "Exit" are present to reset the interface and close the application. On the right, performance metrics for PSNR, MSE, and the number of shifts are displayed to evaluate the steganography process.

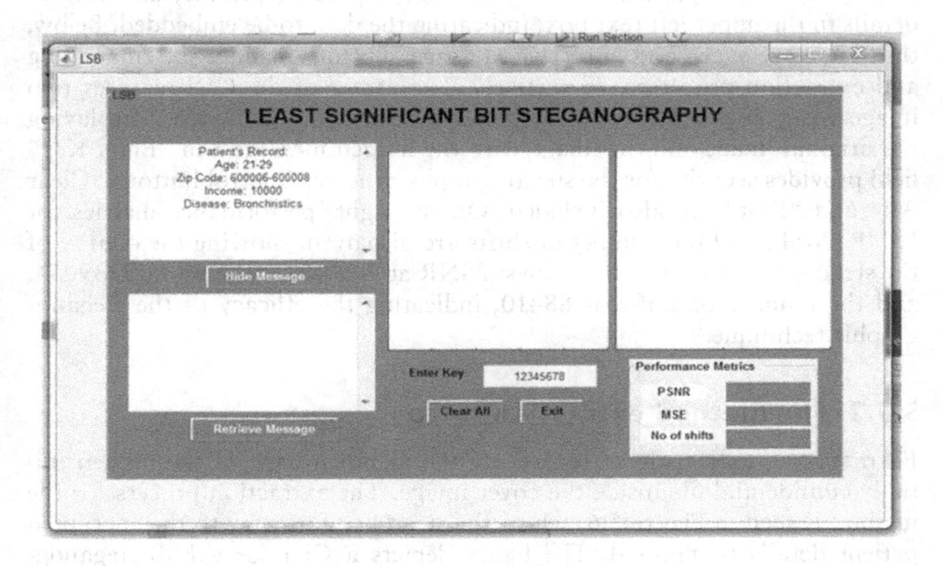

Figure 5.4 Interface depicting the message and stego key.

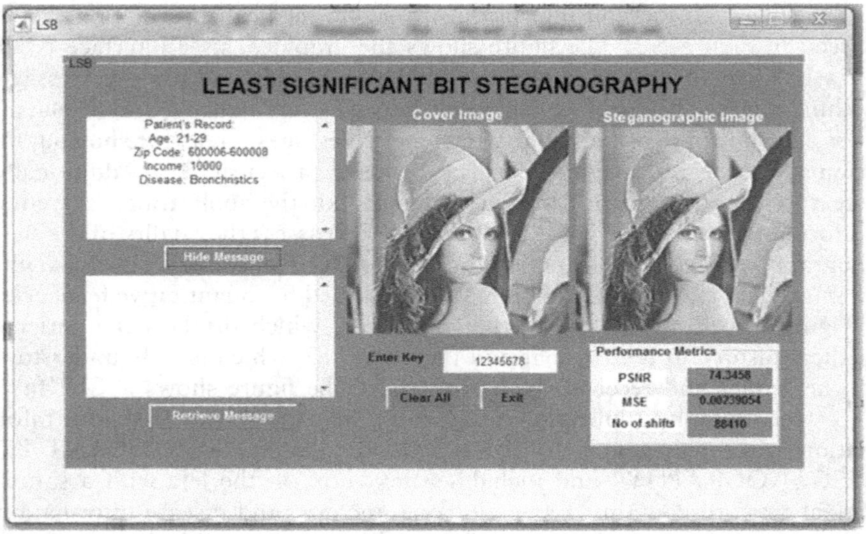

Figure 5.5 Interface of the embedding result.

Figure 5.5 depicts the interface that displays the system's embedding stage result. This indicates the original cover image as well as the steganographic image. The figure illustrates a GUI for a LSB steganography application, showcasing the process of hiding and retrieving sensitive information within images. The interface, "LEAST SIGNIFICANT BIT STEGANOGRAPHY," displays a sample patient's record with age, zip code, income, and disease details in the upper left text box, indicating the data to be embedded. Below, the "Hide Message" and "Retrieve Message" buttons manage the embedding and extraction processes, respectively. The center of the GUI features two image areas labeled "Cover Image" and "Steganographic Image," displaying the original image and the image with the hidden message. An "Enter Key" field provides security for the steganography process. Control buttons "Clear All" and "Exit" are also included. On the right, performance metrics for PSNR, MSE, and the number of shifts are displayed, showing the quality of the stego image with specific values: PSNR at 74.3458, MSE at 0.00239054, and the number of shifts at 88410, indicating the efficacy of the steganographic technique.

5.3.2 The method extraction stage

Extraction is extracting or recovering the identical copy of the hidden initially confidential file inside the cover image. The extraction process for the method is seen in Figure 5.6, where the stego key for entry to the encrypted patient details is inputted. The figure depicts a GUI for a LSB steganography application. This interface, labeled "LEAST SIGNIFICANT BIT

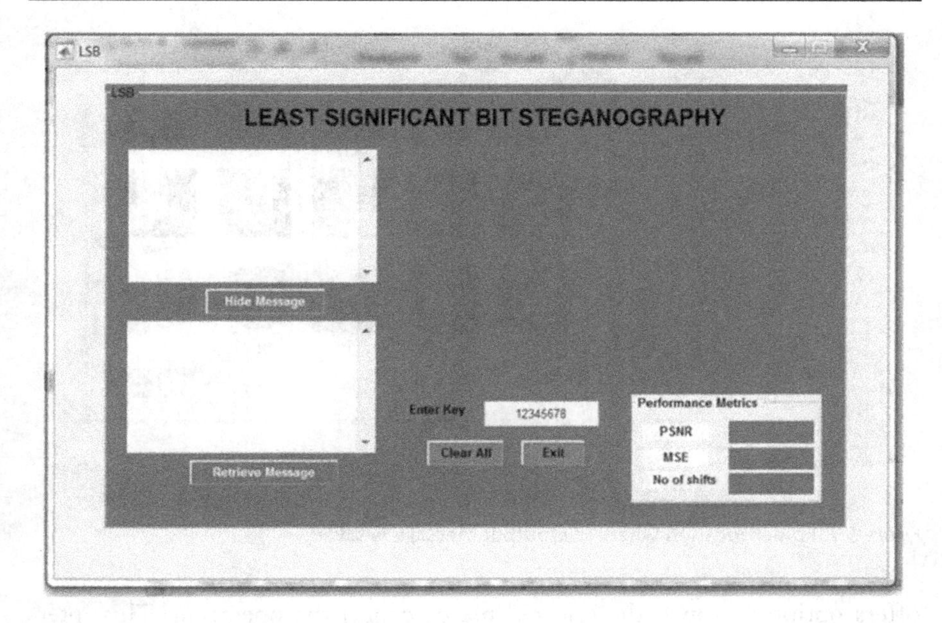

Figure 5.6 Method extraction interface.

STEGANOGRAPHY," is designed for embedding and extracting hidden messages within images. It features two main text boxes on the left: one for entering the message to be hidden and the other for displaying the retrieved message. Below these text boxes are "Hide Message" and "Retrieve Message" buttons that initiate the embedding and extraction processes. In the center, an "Enter Key" field allows the user to input a security key required for the steganography operations. The GUI also includes control buttons for clearing all inputs ("Clear All") and exiting the application ("Exit"). On the right-hand side, performance metrics fields are provided to display the PSNR, MSE, and the number of shifts, which are used to assess the quality and effectiveness of the steganographic process.

The user first type in the stego key (secured key) and then selects the stego image from the document saved in the computer or external storage gadgets, as shown in Figure 5.7. The location where the output message is saved and selected to be uploaded is shown in Figure 5.7. The figure displays a file selection dialog box within the MATLAB environment, specifically designed for choosing a steganographic image to be used in the LSB steganography application. The dialog box "Select STEGANOGRAPHIC IMAGE" shows the contents of the "LSB" folder, which includes various image files with different extensions such as .jpg, .png, .jpeg, and .bmp. The images displayed include standard test images like "lena," and medical images labeled "mdb" series, as well as other assorted images like "UNILORN" and "walkway." The selected file, highlighted in the dialog, is named "sample1," and the dialog

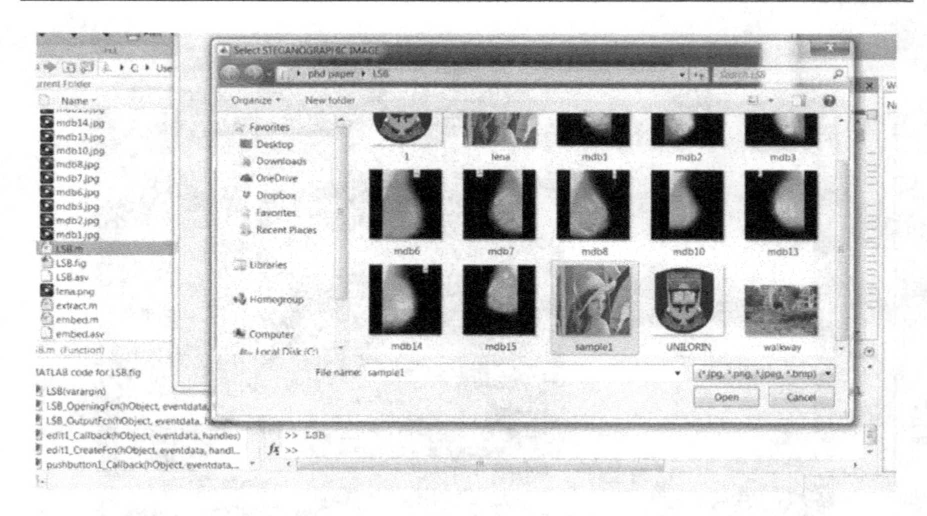

Figure 5.7 Folder location where the output message is saved.

offers options to open the selected file or cancel the operation. This interface facilitates the user's ability to browse and choose the appropriate image for embedding or retrieving hidden messages, seamlessly integrating with the steganography application's workflow.

The application output is shown in Figure 5.8, displaying the secured patient medical information and the original cover image. The concealed information

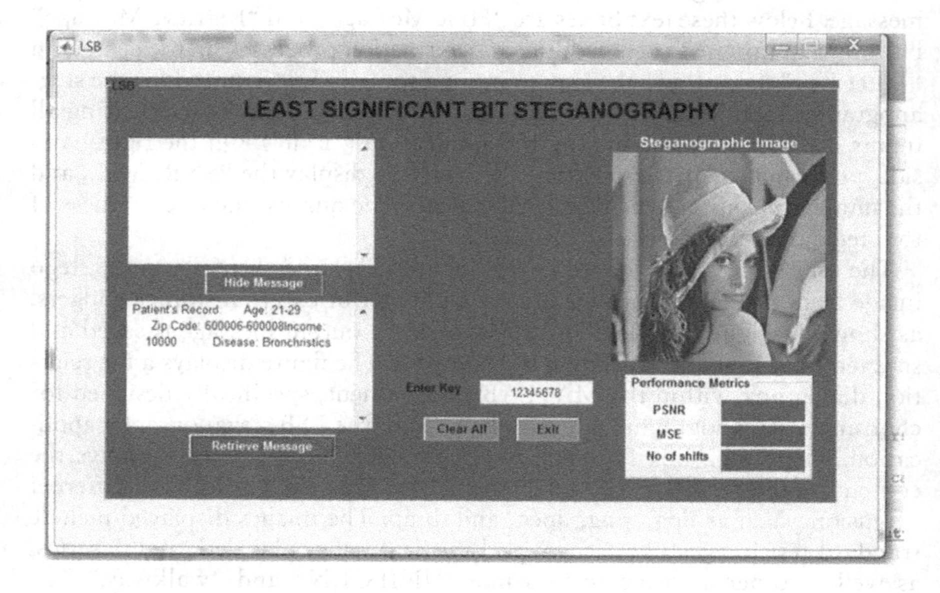

Figure 5.8 Extraction interface.

is seen in the retrieve message and the original cover message in Figure 5.8, which depicts the proposed system's extraction process. The figure shows a GUI for a LSB steganography application designed to embed and retrieve messages within images. The GUI, titled "LEAST SIGNIFICANT BIT STEGANOGRAPHY," includes text boxes for message input and retrieval. The lower left text box displays a sample patient's record with demographic and health information, ready to be hidden in an image. The "Hide Message" and "Retrieve Message" buttons facilitate embedding and extracting the message. The right-hand side of the GUI displays the "Steganographic Image" section, which shows the image where the message has been embedded. An "Enter Key" field provides an added security measure for the process. Control buttons "Clear All" and "Exit" allow users to reset the interface and close the application. Additionally, performance metrics for PSNR, MSE, and the number of shifts are included but currently not populated, providing feedback on the quality of the steganographic process once it is executed.

5.3.3 Performance analysis

The PSNR and MSE are the primary metrics used to examine steganography system performance. The modified LSB PSNR and MSE values were likened to the standard LSB. The formula for the two metrics is shown below: the investigation assesses the method by calculating the PSNR and MSE [36–39].

i. PSNR is calculated by employing Equation (5.1):

$$PSNR = 10\log\left(\frac{255^{\,2}}{MSE}\right) \tag{5.1}$$

ii. MSE is calculated by employing Equation (5.2):

$$MSE = \frac{1}{mn}\sum_{0}^{m-1}\sum_{0}^{n-1} \| c - s \|^2 \tag{5.2}$$

Table 5.1 presents the performance metrics of a LSB steganography application, evaluating four different samples. Each sample varies by the number

Table 5.1 Assessment of the suggest method

Samples	*Number of stego key characters*	*PSNR*	*MSE*	*No. of shifts*
1	8	74.3458	0.002391	88,410
2	5	75.0789	0.002019	32,640
3	8	79.1473	0.000791	88,410
4	5	80.364	0.000598	32,640

of stego key characters embedding the message. The metrics include PSNR, MSE, and the number of shifts required for the embedding process. Sample 1, using an 8-character stego key, has a PSNR of 74.3458, an MSE of 0.002391, and 88,410 shifts. Sample 2, with a 5-character key, shows improved metrics with a PSNR of 75.0789, an MSE of 0.002019, and 32,640 shifts. Sample 3, also with an 8-character key, exhibits the highest PSNR of 79.1473 and a lower MSE of 0.000791, maintaining the same number of shifts as Sample 1. Sample 4, with a 5-character key, achieves the best overall performance with a PSNR of 80.364, an MSE of 0.000598, and 32,640 shifts. These results indicate that using fewer stego key characters can enhance the quality of the stego image, as evidenced by higher PSNR and lower MSE values, while reducing the number of shifts required.

It is noticed from Figure 5.9a and b that there is no variance in both the concealment object and steganographic object (image with medical information

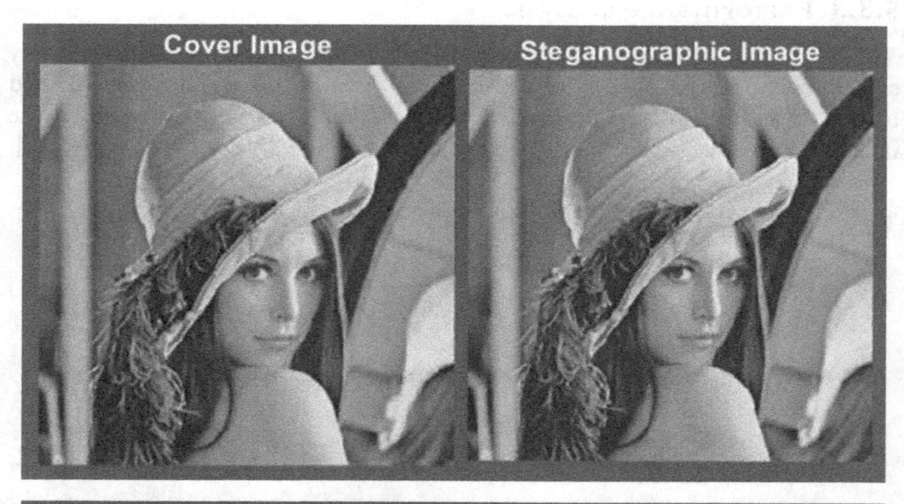

Figure 5.9 (a,b) Image of original concealment object and steganographic object.

hidden in it). Figure 5.9b presents two images: the "Cover Image" on the left and the "Steganographic Image" on the right. Both images depict a similar scene of a fountain with a university emblem in the background, surrounded by greenery and buildings. Upon close inspection, it is apparent that there is no visible difference between the two images, even though the "Steganographic Image" has hidden medical information embedded within it using the LSB steganography technique. This lack of visible variance demonstrates the effectiveness of the LSB method in concealing data without altering the visual appearance of the cover image, thereby maintaining its integrity and making the hidden information undetectable to the naked eye. The seamless integration of secret data in the steganographic image highlights the technique's robustness in preserving the original image's visual quality while securely embedding sensitive information.

This deduced that the modified LSB steganography system is more robust and is invisible to standard LSB. It can also be seen from Table 5.2 that the proposed method, which is the modified LSB technique, performed better than the existing one because it has a higher PSNR of 80.5 and lower MSE of 0.000598 when likened to state of the art. The table compares the performance metrics of various LSB steganography systems as evaluated by different authors, including Devi [40], Dewangga et al. [41], Setyono and Setiadi [42], Habeeb [23], and the proposed system from 2020. The key metrics compared are PSNR and MSE, which are used to assess the quality and imperceptibility of the stego images. Devi's [40] work reports PSNR values ranging

Table 5.2 Assessment with state-of-the-art LSB

Authors	*Type of cover image*	*PSNR*	*MSE*
Devi [40]	Image	61.4733 67.6697	0.0463 0.0210
Dewangga et al. [41]	Image	59.8584 60.2599 60.6315 63.3499 62.2151	0.0677 0.0617 0.0567 0.0303 0.0394
Setyono and Setiadi [42]	Image	63.3908 62.6165 63.5195 63.2185	0.0298 0.0340 0.0289 0.0310
Habeeb [23]	Image	45.23 43.59 50.80 55.35	They didn't evaluate their work with MSE
Proposed system, 2020	Image	74.3458 75.0789 79.1473 80.364	0.002391 0.002019 0.000791 0.000598

from 61.4733 to 67.6697 and MSE values from 0.0463 to 0.0210. Dewangga et al. [41] present a series of PSNR values between 59.8584 and 63.3499 with corresponding MSE values from 0.0677 to 0.0303. Setyono and Setiadi [42] show PSNR values from 62.6165 to 63.5195 and MSE values between 0.0340 and 0.0289. Habeeb's [23] study reports PSNR values from 43.59 to 55.35 but does not evaluate MSE. In contrast, the proposed system from 2020 demonstrates significantly higher PSNR values, ranging from 74.3458 to 80.364, and much lower MSE values, from 0.002391 to 0.000598. The system also used the number of shifts as an additional performance metric, which no previous researcher had used. This indicates that the proposed system achieves superior image quality and better imperceptibility of the hidden information compared to the other methods reviewed.

5.4 CONCLUSIONS

A modified LSB image steganography method was proposed and evaluated using key metrics such as PSNR, MSE, and the number of shifts. The prototype for this approach was successfully implemented and rigorously tested. Results indicated that employing more bits in the process increases the risk of altering the image, thereby reducing the perceptibility and detectability of the hidden information. The modified LSB method demonstrated superior performance to traditional LSB methods, with PSNR values ranging from 74.3458 to 80.364, indicating higher image quality and less distortion. MSE values ranged from 0.002391 to 0.000598, showcasing minimal error and high fidelity of the stego images. The numerical analysis confirmed that the proposed system produced significantly more secure stego photos, making it difficult for intruders to detect an attack. Additionally, it was observed that a longer stego key enhances system security, with the number of shifts ranging from 32,640 to 88,410, further improving protection. This study contributes to the body of knowledge by presenting a modified LSB method that outperforms previous versions regarding robustness, image quality, and security, emphasizing the importance of the number of shifts in enhancing the security of LSB steganography techniques.

REFERENCES

1. Internet: Electronic Health Record, http://en.wikipedia.org/wiki/Electronic_health_record (2015, April 23).
2. Internet: Electronic Health Records Overview, http://www.himss.org/files/HIMSSorg/content/files/Code%20180%20MITRE%20Key%20Components%20of%20an%20EHR.pdf (2015, April 23). http://dx.doi.org/10.1016/j.compeleceng.2017.08.020
3. A. Oluwakemi Christiana, A. Noah Oluwatobi, G. Ayomide Victory, and O. Roseline Oluwaseun, "A secured one time password authentication technique using (3, 3) visual cryptography scheme," *Journal of Physics: Conference Series*, 1299, no. 1, 2019

4. N. O. Akande, C. O. Abikoye, M. O. Adebiyi, A. A. Kayode, A. A. Adegun, and R. O. Ogundokun, "Electronic medical information encryption using modified blowfish algorithm," *Lecture Notes in Computer Science (including subseries Lecture Notes in Artificial Intelligence and Lecture Notes in Bioinformatics)* 11623 LNCS, pp.166, 2019.
5. A. A. Emmanuel, O. Mukaila, A. M. Olubunmi, O. O. Roseline, L. A. Folaranmi, A. A. Elizabeth, A. P. Ojochenemi, and E. P. Anyaiwe, "Vehicle-caused road accidents of four major cities in north-central region of Nigeria. (2010–2015)," *International Journal of Civil Engineering and Technology*, vol. 10, no. 2, p.124, 2019.
6. J. Liao, S. Yin, X. Guo, A. K. Li, and Sangaiah, "Medical JPEG image steganography based on preserving inter-block dependencies," *Computers & Electrical Engineering*, vol. 67, pp. 320–329, 2018.
7. P. Malathi, M. Manoaj, R. Manoj, V. Raghavan, and R. E. Vinodhini, "Highly improved DNA based steganography," *Procedia Computer Science*, vol. 115, pp.651–659, 2017.
8. M. Hussain, A. Wahab, W. A. Idris, I. B. Ho, and K. H. Jung, "Image steganography in spatial domain: A survey," *Signal Processing: Image Communication*, vol. 65, pp.46–66, 2018.
9. H. Al-Dmour and A. Al-Ani, "A steganography embedding method based on edge identification and XOR coding," *Expert Systems with Applications*, vol. 46, pp.293–306, 2016.
10. R. Tavoli, M. Bakhshi, and F. Salehian, "A new method for text hiding in the image by using LSB," *International Journal of Advanced Computer Science and Applications*, vol. 7, no. 4, pp.126–132, 2016.
11. B. Datta, U. Mukherjee, and S. K. Bandyopadhyay, "LSB layer independent robust steganography using binary addition," *Procedia Computer Science*, vol. 85, pp.425–432, 2016.
12. A. Oluwakemi Christiana, A. Noah Oluwatobi, G. Ayomide Victory, and O. Roseline Oluwaseun, "A secured one time password authentication technique using (3, 3) visual cryptography scheme," *Journal of Physics: Conference Series*, vol. 1299, no. 1, p. 012059, 2019.
13. O. N. Akande, O. C. Abikoye, A. A. Kayode, O. T. Aro, and O. R. Ogundokun, "A dynamic round triple data encryption standard cryptographic technique for data security," *Lecture Notes in Computer Science (including subseries Lecture Notes in Artificial Intelligence and Lecture Notes in Bioinformatics)*, 12254 LNCS, pp. 487–499, 2020.
14. N. O. Akande, C. O. Abikoye, M. O. Adebiyi, A. A. Kayode, A. A. Adegun, and R. O. Ogundokun, "Electronic medical information encryption using modified blowfish algorithm," *Lecture Notes in Computer Science (including subseries Lecture Notes in Artificial Intelligence and Lecture Notes in Bioinformatics)*, 11623 LNCS, pp. 166–179, 2019.
15. R. O. Ogundokun, O. C. Abikoye, S. Misra, and J. B. Awotunde, "Modified least significant bit technique for securing medical images," *Lecture Notes in Business Information Processing*, vol. 402, pp. 553–565, 2020.
16. M. M. Amin, M. Salleh, S. Ibrahim, M. R. Katmin, and M. Z. I. Shamsuddin, "Information hiding using steganography," *In 4th National Conference of Telecommunication Technology, 2003. NCTT 2003 Proceedings*, pp. 21–25. IEEE, 2003.
17. C. K. Chan and L. M. Cheng, "Hiding data in images by simple LSB substitution," *Pattern Recognition*, vol. 37, no. 3, pp.469–474, 2004.
18. D. C. Wu and W. H. Tsai, "A steganographic method for images by pixel-value differencing," *Pattern Recognition Letters*, vol. 24, no. 9–10, pp. 1613–1626, 2003.

19. C. M. Wang, N. I. Wu, C. S. Tsai, and M. S. Hwang, "A high-quality steganographic method with pixel-value differencing and modulus function," *Journal of Systems and Software*, vol. 81, no. 1, pp.150–158, 2008.
20. O. C. Abikoye, U. A. Ojo, J. B. Awotunde, and R. O. Ogundokun, "A safe and secured iris template using steganography and cryptography," *Multimedia Tools and Applications*, vol. 79, no. 31–32, pp.23483–23506, 2020.
21. G. M. Kamau, "An enhanced least significant bit steganographic method for information hiding" (Doctoral dissertation), 2014.
22. N. Provos and P. Honeyman, "Detecting steganographic content on the internet," Center for Information Technology Integration, 2001.
23. A. Habeeb, "A new method for hiding text in a digital image," *Journal of Southwest Jiaotong University*, vol. 55, no. 2, 2020. doi: 10.35741/issn.0258-2724.55.2.4
24. N. Masoud and I. Ghazi, "A modification of least significant digit (LSD) digital watermark technique," *International Journal of Computer Applications*, vol. 179, pp.4–6, 2018.
25. T. Ali and A. Doegar, "A novel approach of LSB based steganography using parity checker," *International Journal of Advanced Research in Computer Science and Software Engineering*, vol. 5, no. 1, 2015
26. M. Juneja and P. S. Sandhu, "Improved LSB-based steganography techniques for color images in spatial domain," *IJ Network Security*, vol. 16, no. 6, pp.452–462, 2014.
27. M. Juneja and P. S. Sandhu, "An improved LSB based steganography technique for RGB colour images," *International Journal of Computer and Communication Engineering*, vol. 2, no. 4, p.513, 2013.
28. S. Gupta, G. Gujral, and N. Aggarwal, "Enhanced least significant bit algorithm for image steganography," *IJCEM International Journal of Computational Engineering & Management*, vol. 15, no. 4, pp.40–42, 2012.
29. G. Chawla, R. Yadav, and R. A. Saini, "Novel approach for image steganography in spatial domain using last two bits of pixel value," *International Journal of Security*, vol. 5, no. 2, pp.51–61, 2012.
30. A. K. Sahu and G. Swain, "Reversible image steganography using dual-layer LSB matching," *Sensing and Imaging*, vol. 21, no. 1, p.1, 2020.
31. H. Al-Dmour and A. Al-Ani, "Quality optimized medical image information hiding algorithm that employs edge detection and data coding," *Computer Methods and Programs in Biomedicine*, vol. 127, pp.24–43, 2016.
32. P. Asmitha, C. Rupa, S. Nikitha, J. Hemalatha, and A. K. Sahu (2024). "Improved multiview biometric object detection for anti spoofing frauds," Multimedia Tools and Applications, pp.1–17, 2024.
33. A. K. Sahu, K. Umachandran, V. D. Biradar, O. Comfort, V. Sri Vigna Hema, F. Odimegwu, and M. A. Saifullah, "A study on content tampering in multimedia watermarking," *SN Computer Science*, vol. 4, no. 3, p.222, 2023.
34. S. Kamil Khudhair, M. Sahu, and A. K. Sahu, "Secure reversible data hiding using block-wise histogram shifting," *Electronics*, vol. 12, no. 5, p.1222, 2023.
35. A. K. Sahu, "A logistic map-based blind and fragile watermarking for tamper detection and localization in images," *Journal of Ambient Intelligence and Humanized Computing*, vol. 13, no. 8, pp.3869–3881, 2022.
36. K. R. Raghunandan, R. Dodmane, K. Bhavya, N. K. Rao, and A. K. Sahu, "Chaotic-map based encryption for 3D point and 3D mesh fog data in edge computing," *IEEE Access*, vol. 11, pp.3545–3554, 2022.
37. S. Deb, A. Das, and N. Kar, "An applied image cryptosystem on Moore's automaton operating on δ(qk)/𝔽2," *ACM Transactions on Multimedia Computing, Communications and Applications*, vol. 20, no. 2, pp.1–20, 2023.

38. S. Deb, S. Pal, and B. Bhuyan, "NMRMG: Nonlinear multiple-recursive matrix generator design approaches and its randomness analysis," *Wireless Personal Communications*, vol. 125, no. 1, pp.577–597, 2022.
39. K. S. Roy, S. Deb, and H. K. Kalita, "A novel hybrid authentication protocol utilizing lattice-based cryptography for IoT devices in fog networks," *Digital Communications and Networks*, vol. 10, no. 1, 2022.
40. K. J. Devi, "A secure image steganography using LSB technique and pseudo-random encoding technique," Project thesis submitted in the Department of Computer Science and Engineering, National Institute of Technology Rourkela, India, 2013.
41. A. P. Dewangga, T. W. Purboyo, and R. A. Nugrahaeni, "A new approach of data hiding in BMP images using LSB steganography and Caesar Vigenere cipher cryptography," *International Journal of Applied Engineering Research*, vol. 12, no. 3, pp.10626–10636, 2017.
42. A. Setyono and R. I. M. Setiadi, "Securing and hiding secret message in image using XOR transposition encryption and LSB method," *IOP Conference Series: Journal of Physics*, vol. 1196, 2019. doi:10.1088/1742-6596/1196/1/012039

Chapter 6

Multimedia security

Basics, its subfields, and allied areas

P. K. Paul, Ritam Chatterjee, Nilanjan Das, Sushil K. Sharma, R. Saavedra, and Abhijit Bandyopadhyay

6.1 INTRODUCTION

Multimedia is a valuable component of information technology (IT). Multimedia security is a vast area as multimedia denotes many things such as digital watermarking, information, and data encryption, authentication of the multimedia, steganography, digital rights management (DRM), etc. (refer to Figure 6.1). Therefore, security aspects also pertain to all the mentioned areas. It is a fact that wire/wireless systems offer multimedia services, and there may be a case of transmission errors [1, 2]. Watermarking and its allied concerns are highly associated with robustness regarding data transmission, which are the messages categorized as "watermarked multimedia."

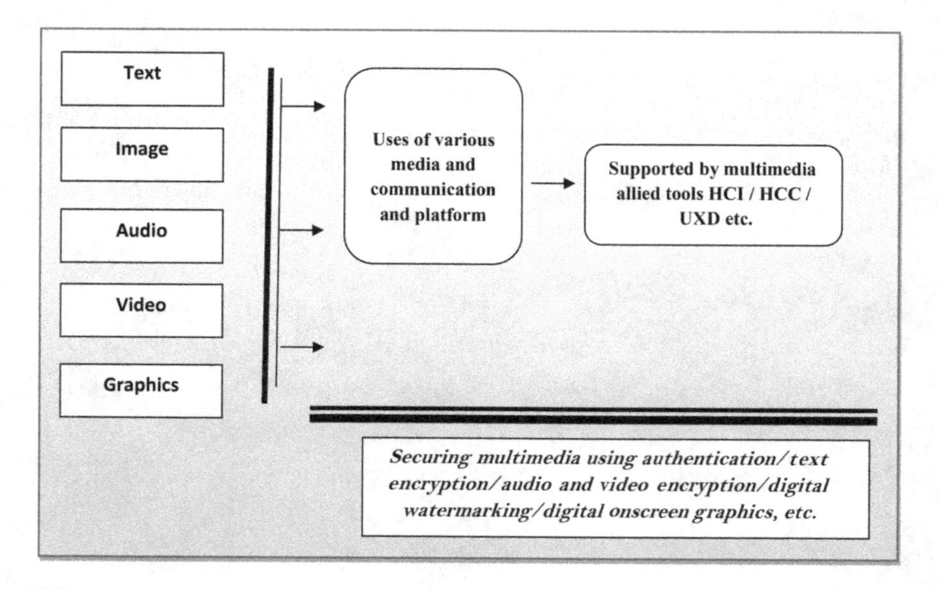

Figure 6.1 Traditional multimedia systems and securing components.

 DOI: 10.1201/9781032663647-6

Multiple description coding (MDC) is a handsome mechanism dedicated to transferring information to reliable and concerned networks that are over non-prioritized as well as treated as unpredictable networks. MDC is extensively studied and it is applicable to audio and video, documents and images, including reliable video sources, and all these can be considered under the source coding scheme. MDC can be considered as a coding that is effectively useful in proper multimedia watermarking, and such MDC schemes have emerged in the recent past. Such MDC schemes have exhibited some advantages. Securing multimedia products by proper protection with rights management and the newest technologies are applicable in proper multimedia processing as well as communications. As multimedia services are applicable in diverse places, it is also essential for the purpose of ensuring confidentiality and integrity in regard to multimedia data. Digital media and its security management aspects and techniques are also increasing rapidly for advanced malware attacks, and though the existing works are initially dedicated to proper and effective encryption techniques to provide multimedia security, old technologies may be replaced whenever required and possible [3–5].

Multimedia products can be of different types and may vary from situation to situation based on the context. In general, the broad multimedia categories include linear and nonlinear multimedia. Linear multimedia is not so interactive, and users have less control over this technology [6–8]. On the other hand, nonlinear multimedia is more interactive and user-friendly, and users may have more navigation control. Some of the features of linear multimedia areas are as follows:

- Linear multimedia is not so interactive, and users have less control over this technology.
- In linear communication, content may be shared, and the users have no such role.
- Less multimedia enriched.
- Little control over the product, tools, and presentation.
- Still images or a simple video is an example of linear multimedia.

As far as nonlinear multimedia is concerned, the following are some of the essential attributes and features:

- Nonlinear multimedia is more interactive and user-friendly; users may have more navigation control.
- The complex information may be shared using nonlinear multimedia.
- Massive information highway management becomes possible using nonlinear multimedia.
- Games, virtual reality, and augmented reality are examples of nonlinear multimedia.

6.2 OBJECTIVE OF THE WORK

The work "Multimedia Security: Foundation, Stakeholders, Importance and Some Allied Aspects: An Overview" aimed at the following:

- To know about the multimedia and multimedia security aspects in a fundamental way, including features and attributes.
- To gather data encryption and digital watermarking, such as nature, attributes, and features, briefly.
- To concisely learn the fundamentals of data encryption, including audio and video encryption.
- To briefly discover the aspects and thoughts related to multimedia authentication and DRM.
- Prepare a list of suggestions for building and developing secure and sophisticated multimedia security systems.

6.3 METHODS

The work "Multimedia Security: Basics, Its Subfields and Allied Areas: A Review" is theoretical and wholly dedicated to documenting briefly the multimedia security. In this chapter, existing works related to multimedia security have been referred, analyzed, and reported. Apart from the secondary sources, some primary sources are supposed to find out the contemporary scenario of multimedia security. Further, specific multimedia-related security analyses related to other subfields have also been studied, such as digital watermarking, audio encryption, data encryption, video encryption, multimedia authentication, DRM, etc.

6.3.1 Existing works

This chapter is dedicated to analyzing and preparing concise documentation related to multimedia security, and therefore, among different multimedia security segments, few have been selected and subsequently studied scientifically. Here, some of the works have been reported alphabetically. Cheddad et al. [9] surveyed digital image steganography and analyzed the results using various techniques of steganography that are available presently. Their work followed several general standards and guidelines that were taken from diverse literature. It recommended several suggestions and supported the object-oriented embedding techniques for steganography. It provided various embodied disciplines of information concealment and briefly described steganography. It made a comparison between steganography, watermarking, and encryption. It showed some practical steganography applications. It elaborated on some steganography methods and showed steganography in the image frequency domain. It also analyzed some steganographic tools and tried to find the drawbacks of various existing techniques. Dang and Chau [10] worked on image encryption for a safe and secure method of transmission of multimedia applications over the Internet. Their work was associated with discrete wavelet transform (DWT) for image compression. It also showed the process of blocking the cipher text for image encryption with the help of data encryption standard (DES). It has been shown that the simulated results are responsible for increasing the security of image transmission when transmitted through the Internet. It also showed the process of improving the image's transmission rate when transmitted over the Internet. It worked on image compression and image encryption. For source coding encryption, it used the DWT technique, and for image data encryption, it followed the DES. It showed the single round of DES algorithm. It discussed the results of the experiments. Liu et al. [11] worked on managing digital rights for content distribution. They reviewed the current state of DRM, focusing on security technologies. Their work dealt with various legal implications related to DRM and sought to better understand the current situation of content management on a legal and technical basis. It tried to project prospects of DRM. It briefly describes the DRM and shows a typical DRM model. It analyzed the existing market and the potential markets for DRM. It also demonstrated the different security mechanisms of DRM. It described various tamper resistance systems and the procedures for dealing with other legal issues. It shows customer concerns and mentions various standards for DRM architecture.

Morkel et al. [12] provided an overview of the process of image steganography. Their work described the various usability of image steganography and its different techniques. It depicted the necessity of good image steganographic algorithms and suggested suitable ones for different purposes. It shows multiple categories of steganography. It discussed other image and transform domains and evaluated varying steganography techniques. It also compared various image steganography algorithms. Prajwalasimha and Surendra [13] worked on discrete dyadic transformation on multimedia data encryption. Their work proposed a confusion and diffusion process based on a private key encryption algorithm. It used neighboring pixels that were tightly correlated with each other to be considered for multimedia data processing.

To randomize the pixel positions, it used discrete dyadic transformation in the confusion algorithm, and to change the value of each pixel, it used the modified Ginger Breadman sequence generator process in the diffusion algorithm. It considered 128 bits of key length. It proposed a cryptosystem with a flow diagram. It discussed the result of the experiment. It compared the average correlation, entropy, and mean square error for the encrypted and standard images. For the classic images, the result of the analytical test performance of the proposed algorithm has been shown. It also showed the computational time for standard images the proposed algorithm took. Stanton et al. [14] surveyed techniques for storing multimedia data safely and securely, emphasizing data encryption. Their work included theoretical approaches and prototype systems. It discussed the protection of multimedia data in storage. It overviews various systems' significant features and suggests the most appropriate solution. It gave a set of conditions for assessing a storage solution based on confidentiality, integrity, availability, and the system's performance. It compared the main differences between the systems and tried to identify various features of the storage systems. It also discussed some practical applications of storage devices that were helpful in multimedia environments. It discussed different survey techniques: NASD (Network Attached Secure Disks), Self Securing Storage (S4), SFS-RO (Secure File System-Read Only), PLUTUS, and SiRiUS (Securing Remote Untrusted Storage). It classified and compared various storage systems. It provides different applications for multimedia system mapping.

Su et. al. [15] surveyed chaos-based encryption technology in a multimedia security context. It analyzed the requirements of multimedia encryption and discussed the evaluation methods of multimedia encryption techniques. It assessed the security analysis, time analysis, compression ratio test, and error robustness test. It proposed chaos-based image encryption algorithms. It discussed the entire encryption process and partial encryption process for an image and also made a comparison of performance. Encryption of the raw video data, encryption of the video data in the compression process, and the compressed video data have also been discussed. It also discussed chaos-based audio encryption algorithms with the process of full encryption and partial encryption and compared performance. Subramanya and Yi [16] worked on managing digital rights. It gave a brief description of DRM and the major functionalities of DRM. The chapter diagrammatically shows a comprehensive overview of the flow of content from creator to consumer. High-level architecture and leading components of an ideal DRM system were also discussed. It discussed content delivery, security and content protection, metadata, rights object and usage rights, and license generation. It made a comprehensive outline of the operation of a DRM system. It diagrammatically shows an overview of principal operations in an ideal DRM system.

6.4 MULTIMEDIA AND MULTIMEDIA SECURITY: FUNDAMENTALS

Multimedia is a part of IT and is an emerging application area of the IT and computing industry. Multimedia combines different kinds of media for audiovisual or textual communication. In multimedia, standard media are e-media (online and offline) of other file formats. Multimedia is featured with varying forms of content, which may help in interactive communication. Multimedia was previously considered a system, but gradually it has developed as a field of study and technology [17, 18]. Today, many universities and higher educational institutes have added multimedia to their course catalog. Below are brief features of multimedia:

- Multimedia is a composition of different kinds of media.
- The sender and receiver are necessary for any communication, including multimedia-based communication.
- Text, audio, video, graphics, and other file formats are considered important content in multimedia.
- Multimedia is a branch of IT and other electronics systems.
- Multimedia may be static or dynamic, and it may also be interactive.
- Multimedia products are developed in different stages using various tools, technologies, and systems.
- Multimedia may be considered not only a subject but also a technology and field of study.
- Multimedia depends on hardware, software, database systems, and other allied technologies [5].
- Multimedia needs the support of various latest technologies such as human–computer interaction (HCI), usability engineering, UXD, etc.
- Multimedia products may help navigate hyperlink-related activities.

Multimedia security is a concept, tool, and mechanism ultimately dedicated to healthy and proper content-based protection [5, 8]. As far as information management is concerned, content is considered a higher level of representation dedicated to creating, processing, and storing the semantics of the data. Therefore, content may be integrated and made available in the following diverse formats:

- Audio and sounds
- Images
- Video and allied visuals
- Text and contents
- Graphics

Therefore, the information characteristics may differ depending upon the requirement and naturally the multimedia security-related formats and sources [19]. Here, proper bit rate, including the amount of acceptable error, and display requirements may be considered significant constraints in good content protection.

6.4.1 Attributes and characteristics of multimedia security

Multimedia security usually manages and tackles the following attributes:

- Multimedia security is highly required for sensitive content to offer efficient security that can address more sophisticated security systems to the format conversion or fundamentally from the error of the

communication. Here, in this context, a proper example may be the use of the digital signature in which high-level features are represented, and all these are dedicated to appropriate authentication [20–28].
- Security processing for digital watermarking is integrated with proper signal processing activities, ultimately handling reasonable bit rate fluctuations. Finally, digital watermarking and signals can reuse the processing blocks for efficient and adequately structured methods. It is eventually dedicated to attached security processing to verify the content with appropriate bandwidth [6, 19].
- Semantic information is associated with proper and effective content security mechanisms.

It is important to note that multimedia security applications for different purposes are drastically changing with proper development, including DRM. Here are details of various security aspects, techniques, and procedures (refer to Sections 6.4.2–6.4.10).

6.4.2 Data encryption

Data encryption is a method to secure the data. Data encryption encodes the information by following some rules and sending it to others. The receiver receives the encoded information and decodes the information by following some rules. After decoding the data, the receiver gets the original information sent by the sender. It is a secure mechanism to protect the data from unauthorized access. No person can read the encrypted data without decrypting it. So the data is protected from unwanted access. Encrypted data are also known as cipher text. It has two types of keys: encryption key and decryption key. Encryption keys are used to encrypt the data, and the decryption keys are used to decode the data [4, 20]. It provides many benefits to the user. It enhances data security, secures personal information, protects data from unauthorized access, and maintains data privacy and integrity. Figure 6.2 depicts the process and stages of encryption and decryption diagrammatically.

Data encryption is the way of data protection from an unauthorized access during transmission over network. Plenty of techniques are applied to protect information before data transmission over network. Authentic data need to transmit from source to destination by maintaining integrity and originality of data. Data encryption is an effective technique of providing data security. Encryption provides the hiding techniques to conceal original data inside text, or multimedia content. Plaintext is the original data and transformed data is referred to as cyphertext. Therefore, data is encrypted using text or character substitution techniques, or data is hidden by inserting multimedia content.

Due to security gaps in devices, computers, and networks, anonymous persons get access of data as well as other contents, namely, information by

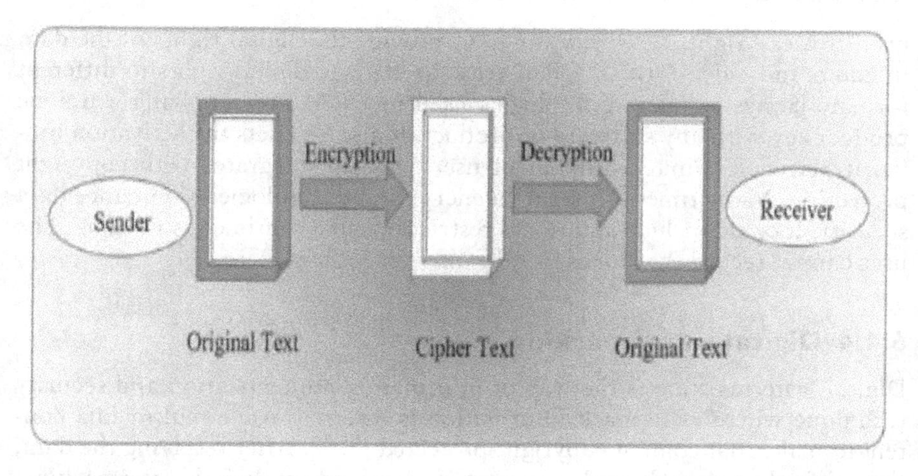

Figure 6.2 The fundamental steps and process of encryption as well as decryption.

cracking security without any barriers. Nowadays, cyberattackers depend on varied types of attacks such as DOS (Denial of Service) attacks, DDOS (Distributed Denial of Service) attacks, or simply Man in Middle attacks, phishing attacks, including malware attacks, password attacks, and crypto-jacking to steal information by breaching security without any concern of authorized person. To protect information from crackers, information must be kept in encrypted form. Therefore, encryption is one of the significant security essential to prevent data hacking. Various encryption methodologies have been followed to protect data.

6.4.3 DRM

DRM is dedicated to managing digital users' rights for the content, and it is further dedicated to linking and protecting users' privileges for different kinds of media in various circumstances. Among the protections and privileges, a few important ones include control viewing, proper duplication of the work, content-related accessibility as well as distribution, etc. [11, 29]. This is very important for balancing and proper information protection, including creating a usability system and a suitable environment development with proper functionality, the cost-effectiveness of the systems, and new digitalization and marketing opportunities. In designing and developing effective business models, DRM's concepts and characteristics are challenging but very important.

DRM is a management procedure that protects the digital rights of any data and ensures the legal access right to any digital content. There are many technologies and tools available to ensure DRM. It provides access control, view control, alteration, modification control, etc. It also ensures unwanted access to the data. Many countries have laws to secure digital rights and protect users' rights. It is very tough to provide the digital rights of the data in audio and video format. DRM tried to manage digital rights in different formats. Some popular technologies used in DRM are supplying a unique product key with any software or product. It also restricts the activation limits. It permits a limited number of users. It also integrates with copyright protection. Many times, it uses data encryption technologies to enhance data security. It could include additional restrictions like runtime restrictions. The user cannot record the videos or take any screenshots.

6.4.4 Digital watermarking

Digital watermarking is the way of information authentication and security technique where watermark information is inserted inside multimedia content to make that content copyright-protected [3, 7]. After receiving the data, the originality of multimedia content is identified by digital watermarking. The main objective of digital watermarking is to authenticate of source of data and protect data using copyright marking [30].

6.4.5 Multimedia authentication

Authentication means validating by following some predefined rules. Multimedia authentication means validating the multimedia data based on some conditions. As the multimedia data are generated electronically, it is very easy to alter the multimedia data. Multimedia content can be of any form: text, image, audio, or video. There are many algorithms for the authentication of multimedia content. Digital watermarking is one of the multimedia authentication methods. It is used to protect the data from unwanted tampering of data. It also ensures the ownership of the data. Some additional digital data has been added to ensure data authentication. Multimedia authentication is mainly used the preserve the original data. It is very useful to protect the right of the owner to any content created by the owner. It is also used for additional security purposes of the data [31, 32]. It can be used to send some hidden data to the user. Multimedia authentication is very much necessary for the multimedia content. Digital signature and media signature are some multimedia authentication techniques. Both are used for multimedia authentication purposes. Multimedia authentication also protects data from data manipulation. It could be easy to detect the manipulated data from the original one. Figure 6.3 shows the multimedia authentication of digital content.

6.4.6 Steganography

Steganography is the technique of hiding data inside any file. Generally, it is used to hide the message from unauthorized access. The data has been embedded into any files like text file, image file, audio file, video file, and many more. Steganography is used to protect the secret data from misuse.

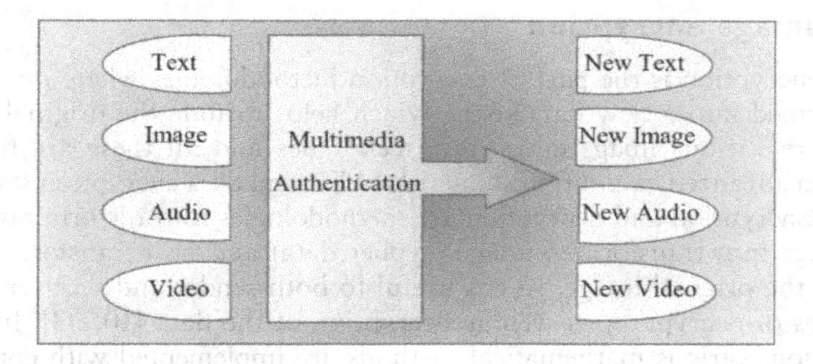

Figure 6.3 The multimedia authentication of digital contents.

Many times the encryption algorithm also combines with the steganography technology. It gives some more security to the data. The secret data can be hidden within any kind of digital multimedia content. When any secret message needs to transmit through the network, then steganography is used. It has many benefits for the user. It increases the security of any data or message. Following the proper algorithm makes it easy to encrypt and decrypt the data [9, 12]. It is also used to track the data. It enhances the security of any files. Steganography technology can be misused. It is one of the cybersecurity threats. The hackers are sending some hidden code with the use of steganography. There is a probability that any virus code can be injected into any computer or mobile phone with the use of steganography. It has many types of steganography algorithms like—steganography related to the text, steganography of the image, steganography of the video and audio, including network steganography. Figure 6.4 shows the message hiding in steganography.

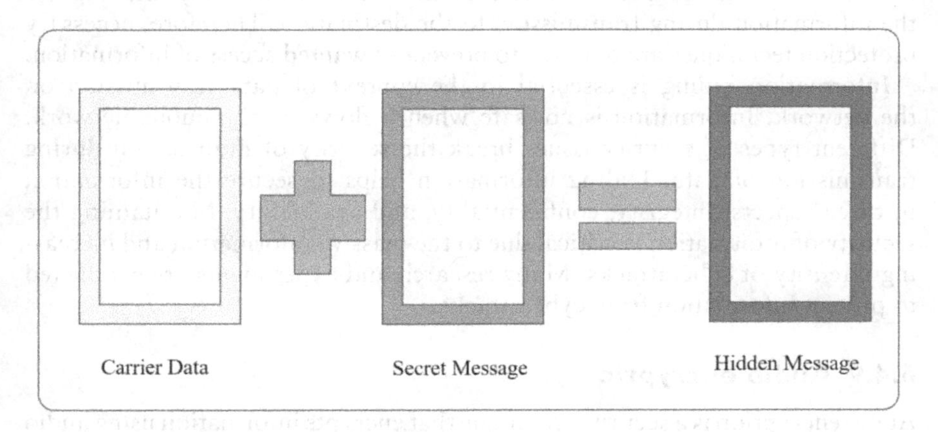

Figure 6.4 Message hiding in steganography.

6.4.7 Image encryption

Image encryption is the part of encryption methodologies where image data transformed into a new data shape which helps to hide the original image. There are various image encryption techniques and all these are followed based on advanced encryption standards (AES) and data encryption standards (DES). Encryption and decryption are methodologies to transform raw original image into transformed image (cypher data) and also transform cypher data to the original image. Key is useful to both sender and receiver for the purposes of encryption as well as decryption of the data [10, 33]. In image encryption, various mathematical methods are implemented with computerized program to alter image pixels and converted into jumbled image which hide original image. Only authentic receivers have authority to decrypt jumbled image (cypher image) with the valid key. The significance of image encryption is increasing nowadays because of the enormous digital images used for different purposes such as images regularly posted in various social media platform, some image published for photography, digital images used for medical purposes, and image captured from satellite or remote sensing devices.

6.4.8 Information hiding

Information is precious in various sectors, including healthcare, education, economy, government, business, social network, agriculture, industry, transportation, and environment. Artificial intelligence, Internet of Things (IoT), fog computing, and cloud and edge computing are the current technologies that produce information in many contexts. For example, real-time information of crop growth status is captured using sensor devices that are part of IoT and transferred to the cloud for further investigation using artificial intelligence to know the present growth of crops. Therefore, massive real-time information is produced and transmitted over the network [15, 34]. There are possibilities that an unauthorized person enter into the network forcefully and capture the information during transmission to the destination. Therefore, necessary protection techniques are required to prevent unwanted access of information.

Information hiding is essential in the context of data transmission on the network. Information is not safe when it flows on the public network. Different types of security issues break the security of information during transmission of data. Hiding information helps to secure the information in three aspects: integrity, confidentiality, and availability. Maintaining the security of information is critical due to the massive information and increasing intensity of cyberattacks. Many research and experiments are conducted to protect information from cyberattacks.

6.4.9 Audio encryption

Audio encryption is a security technique that encrypts information using audio data before sending data to the network. The audio encryption technique is more efficient than other techniques. Information or text data is encrypted in audio data using various methods. Audio encryption techniques are the advanced cryptographic implementation to secure valuable information. In this era, audio encryption is important due to the management of mammoth data produced for different purposes [31, 35]. As the data originated from enormous sources and the format of data is heterogeneous, complexity arises in maintaining data security. Audio encryption helps to overcome the security issues of massive data.

Audio data is converted to a series of binary data. Text data is also converted to binary data to combine with audio binary data to make converted data. On the receiver side, transformed data is decrypted to extract text data. Many audio encryption algorithms are applied to encrypt data, like symmetry-based algorithms, including AES, DES, and RC4, and asymmetry-based algorithms, including RSA and ECC. Various methodologies are used to encrypt audio data to maintain quality and efficiency [36]. LSB, DCT, and DWT are the different ways of audio encryption. Compression, scrambling, and watermarking are the other techniques implemented to enhance data encryption speed, develop strong and complex encryption that creates difficulties in decryption, and maintain data authenticity.

6.4.10 Digital on-screen graphic

On-screen graphics, or digitally originated graphics (DOG), is a graphical watermark image displayed on the portion of videos, live stream videos, and TV channels. On-screen graphics are the same as the logo image displayed on the video screen. This on-screen image is used to authorize videos and protect them from copyright violation. This is one of the authentication techniques to protect and authorize ownership of video content. In TV channels or movies, content with watermark on-screen graphics validates original sources of video content. Various types of DOG are on-screen display, including screen burn-in, clock ident, layout of television news screen, and score bug. The on-screen display settings include volume levels, channel number, and color contrast levels on TV or monitor screen. Screen burn-in is the display problem when the color does not display correctly on the screen, that is, on old CRT monitor [11, 33]. The clock is a digital on-screen graphics that display analogue clock on the screen along with company logo, and this clock is placed on the screen before starting news bulletins and at the end of bulletins. Television news layout is the layout or interface that displays text-based information on the bottom of the video screen in the news channel to broadcast news information. Score bug is the graphical interface displayed on top of the video screen, which is used to display live sports information or any sports result statistics on any sports-related TV channel or in sports-live video streaming [28]. Therefore, digital screen graphics are useful multimedia content that can be applied for many purposes, including maintaining authenticity, integrity, and information.

6.5 CONCLUDING REMARKS

Like multimedia applications and their needs, multimedia security is highly applicable and important in diverse organizations. As massive multimedia-related data, including text, audio, video, images, and graphics, are stored and transmitted in Internet systems, its proper protection is essential. Data misuse, including its tampering and modification for personal purposes, is significant among the security-related aspects and matters. With conventional computer security in modern days, managing all types of content security and diverse semantics of the information is challenging. Thus, proper and effective design and security mechanisms with high bandwidth information management and adequate encryption and authentication functions must be employed. Today, all types of organizations and institutions are highly using digital content. Therefore, its security and mechanisms are also gaining priority in all organizations, namely, private, public, and public–private.

REFERENCES

1. Bhanot, R., & Hans, R. 2015. A review and comparative analysis of various encryption algorithms. International Journal of Security and Its Applications, 9, 289–306.
2. Kadian, P., Arora, S. M., & Arora, N. 2021. Robust digital watermarking techniques for copyright protection of digital data: A survey. Wireless Personal Communications, 118, 3225–3249.
3. Cox, I., Miller, M., Bloom, J., & Honsinger, C. 2002. Digital watermarking. Journal of Electronic Imaging, 11, 414–414.
4. Gai, K., Qiu, M., & Zhao, H. 2017. Privacy-preserving data encryption strategy for big data in mobile cloud computing. IEEE Transactions on Big Data, 7, 678–688.
5. Paul, P. K. 2013. Human computer interaction and its emerging affiliation with information science (IS): An overview. Journal of Business and Management, 5, 162–167.
6. Delp, E. J. 2005. Multimedia security: The 22nd century approach. Multimedia Systems, 11, 95–97.
7. Embaby, A. A., Shalaby, M. A. W., & Elsayed, K. M. 2020. Digital watermarking properties, classification and techniques. International Journal of Engineering and Advanced Technology (IJEAT), 9, 2742–2750.
8. Paul, P. K., Bhuimali, A., & Chatterjee, D. 2017. Human centered computing: Healthy gift of social engineering for promoting digital humanities: A short communication. International Journal of Applied Science and Engineering, 5, 13–18.
9. Cheddad, A., Condell, J., Curran, K., & McKevitt, P. 2010. Digital image steganography: Survey and analysis of current methods. Signal Processing, 90, 727–752.
10. Dang, P. P., & Chau, P. M. 2000. Image encryption for secure internet multimedia applications. IEEE Transactions on Consumer Electronics, 46, 395–403.
11. Liu, Q., Safavi-Naini, R., & Sheppard, N. P. 2003. Digital rights management for content distribution. In Conferences in Research and Practice in Information Technology Series, 34, pp. 49–58.

12. Morkel, T., Eloff, J. H., & Olivier, M. S. 2005. An overview of image steganography. In: Proceedings of the Fifth Annual Information Security South Africa Conference (ISSA 2005), Sandton, South Africa, 1, 1–11.
13. Prajwalasimha, S. N., & Surendra, U. 2017. Multimedia data encryption based on discrete dyadic transformation. In: 2017 International Conference on Signal Processing and Communication (ICSPC) (pp. 492–495). IEEE.
14. Stanton, P., Yurcik, W., & Brumbaugh, L. 2005. Protecting multimedia data in storage: A survey of techniques emphasizing encryption. Proceedings of SPIE, 5682, 18–29
15. Su, Z., Zhang, G., & Jiang, J. 2012. Multimedia security: A survey of chaos-based encryption technology. Multimedia-A Multidisciplinary Approach to Complex Issues. InTechOpen, pp. 99–124.
16. Subramanya, S. R., & Yi, B. K. 2006. Digital rights management. IEEE Potentials, 25, 31–34.
17. Kalaivani, K., & Sivakumar, B. R. 2012. Survey on multimedia data security. International Journal of Modeling and Optimization, 2, 36.
18. Paul, P. K, Aithal, P. S., & Bhimali, A. 2018. Computing academics into new age applied science programs and fields emphasizing trend on animation and multimedia technology: An investigation of Indian private universities. International Journal of Scientific Research in Physics and Applied Sciences, 6, 18–26.
19. Kumari, M., Gupta, S., & Sardana, P. 2017. A survey of image encryption algorithms. 3D research, 8, 1–35.
20. Paar, C., Pelzl, J., Paar, C., & Pelzl, J. 2010. The data encryption standard (DES) and alternatives. Understanding Cryptography: A Textbook for Students and Practitioners, Springer, pp. 55–86.
21. Rose, E. 2011. The phenomenology of on-screen reading: University students' lived experience of digitised text. British Journal of Educational Technology, 42, 515–526.
22. Roy, K. S., Deb, S., & Kalita, H. K. 2022. A novel hybrid authentication protocol utilizing lattice-based cryptography for IoT devices in fog networks. Digital Communications and Networks. https://doi.org/10.1016/j.dcan.2022.12.003
23. Deb, S., Pal, S., & Bhuyan, B. 2022. NMRMG: Nonlinear multiple-recursive matrix generator design approaches and its randomness analysis. Wireless Personal Communications, 125(1), 577–597.
24. Asmitha, P., Rupa, C., Nikitha, S., Hemalatha, J., & Sahu, A. K. 2024. Improved multiview biometric object detection for anti-spoofing frauds. Multimedia Tools and Applications, 1–17. https://doi.org/10.1007/s11042-024-18458-8
25. Sahu, A. K., Umachandran, K., Biradar, V. D., Comfort, O., Sri Vigna Hema, V., Odimegwu, F., & Saifullah, M. A. 2023. A study on content tampering in multimedia watermarking. SN Computer Science, 4(3), 222.
26. Kamil Khudhair, S., Sahu, M., KR, R., & Sahu, A. K. 2023. Secure reversible data hiding using block-wise histogram shifting. Electronics, 12(5), 1222.
27. Raghunandan, K. R., Dodmane, R., Bhavya, K., Rao, N. K., & Sahu, A. K. 2022. Chaotic-map based encryption for 3D point and 3D mesh fog data in edge computing. IEEE Access, 11, 3545–3554.
28. Zhang, Z., Pei, Q., Ma, J., & Yang, L. 2009. Security and trust in digital rights management: A survey. International Journal of Network Security, 9, 247–263.
29. Zhou, L., & Chao, H. C. 2011. Multimedia traffic security architecture for the Internet of things. IEEE Network, 25, 35–40.
30. Mohanarathinam, A., Kamalraj, S., Prasanna Venkatesan, G. K. D., Ravi, R. V., & Manikandababu, C. S. 2020. Digital watermarking techniques for image security: A review. Journal of Ambient Intelligence and Humanized Computing, 11, 3221–3229.

31. Gandhi, R. A., & Gosai, A. M. 2015. A study on current scenario of audio encryption. International Journal of Computer Applications, 116, 110–120.
32. Liu, S., Guo, C., & Sheridan, J. T. 2014. A review of optical image encryption techniques. Optics & Laser Technology, 57, 327–342.
33. Wang, Y., Wong, K. W., Liao, X., & Chen, G. 2011. A new chaos-based fast image encryption algorithm. Applied Soft Computing, 11, 514–522.
34. Shifa, A., Asghar, M. N., & Fleury, M. 2016. Multimedia security perspectives in IoT. In: 2016 Sixth International Conference on Innovative Computing Technology (INTECH) (pp. 550–555). IEEE.
35. Kordov, K. 2019. A novel audio encryption algorithm with permutation-substitution architecture. Electronics, 8, 530.
36. Naskar, P. K., Bhattacharyya, S., & Chaudhuri, A. 2021. An audio encryption based on distinct key blocks along with PWLCM and ECA. Non-linear Dynamics, 103, 2019–2042.

Chapter 7

A comprehensive review on multimedia security through blockchain

Yagnyasenee Sen Gupta

7.1 INTRODUCTION

Multimedia content, comprising images, videos, and audio, has become an integral part of our daily communication and information sharing. However, the increasing reliance on digital multimedia has also raised concerns regarding its security, including issues related to unauthorized access, tampering, and piracy according to Giri et al. (2021). Multimedia data is susceptible to a range of security threats, including unauthorized access, piracy, and manipulation. Multimedia security refers to the protection of multimedia content, which includes various forms of digital media such as images, videos, audio, and other interactive elements. Ensuring the security of multimedia is crucial due to the widespread use of digital media in communication, entertainment, education, and business. Traditional security measures often fall short of providing robust protection against these threats as stated by Gurunathan and Rajagopalan et al. (2020). Highlighting the vulnerabilities associated with traditional multimedia storage and distribution systems, this section explores issues such as data tampering, unauthorized access, and intellectual property theft. It emphasizes the need for a robust security framework to safeguard multimedia content in various domains. This research explores the potential of blockchain technology to enhance the security of multimedia content by establishing a decentralized and tamper-proof framework.

The advent of multimedia-rich applications and platforms has revolutionized the way we create, share, and consume content. However, the inherent vulnerabilities of centralized systems and the ease of unauthorized access have raised concerns regarding the security of multimedia content as in Nie et al. (2020). Blockchain technology, originally designed for secure and transparent financial transactions in cryptocurrencies, has shown great potential in addressing these security concerns. In this chapter, a comprehensive evaluation and analysis of the use of blockchain technology in multimedia security, as well as its possible advantages has been made. The chapter discusses the key principles of blockchain technology and multimedia security, evaluates current issues, and offers adaptive improvements. This study aims to give a thorough knowledge of how blockchain might transform the multimedia

DOI: 10.1201/9781032663647-7

security environment through a close analysis of case studies and real-world deployments.

The chapter has been organized as follows: Section 7.2 gives an overview of multimedia security and blockchain; in Section 7.3, the different works carried out in multimedia security with the help of blockchain have been discussed; the possible solution with an illustration has been provided in Section 7.4; and finally, Section 7.5 concludes the chapter.

7.2 BACKGROUND STUDIES

In this section, the different essential elements of multimedia security, its challenges, along an overview of blockchain are explained in brief.

7.2.1 Key components of multimedia security

i. **Access control**

   Authentication: Verifying the identity of users or devices to ensure that only authorized entities can access multimedia content.

   Authorization: Specifying what actions or operations authorized users or systems are allowed to perform on multimedia files.

   Encryption: Using cryptographic techniques to convert multimedia data into a secure format that can only be deciphered by those with the proper decryption keys. This helps prevent unauthorized parties from viewing or altering the content.

ii. **Watermarking**

   Embedding hidden information, known as watermarks, into multimedia content to prove its authenticity or identify its owner. Watermarking is often used to deter piracy and unauthorized distribution.

iii. **Digital Rights Management (DRM)**

   Implementing systems that control the use of multimedia content and enforce copyright policies. DRM technologies often involve encryption, access controls, and licensing mechanisms.

iv. **Steganography**

   Concealing information within multimedia files without altering their perceptual quality. Steganography aims to hide the presence of information rather than protect the information itself.

v. **Blockchain technology**

   Leveraging blockchain for secure and transparent handling of multimedia content. Blockchain provides an immutable, tamper-proof along with a decentralized ledger, ensuring the integrity and provenance of multimedia files.

vi. **Secure transmission**

   Implementing secure communication channels, such as transport layer security (TLS) or secure sockets layer (SSL) to protect multimedia content at the time of transmission over networks.

vii. **Forensics**
Utilizing digital forensics techniques to investigate and analyze incidents of unauthorized access, tampering, or distribution of multimedia content. This involves the recovery and analysis of digital evidence.

viii. **Biometric security**
Integrating different authentication methods based on biometric parameters, such as fingerprint scanning or facial recognition, to improve the security of multimedia devices and systems.

ix. **Security policies and training**
Establishing and enforcing security policies related to the creation, storage, and sharing of multimedia content. Providing training to users on best practices for multimedia security.

7.2.2 Challenges in multimedia security

Rapid technological advancements: Keeping up with evolving multimedia technologies and ensuring security measures are up-to-date.

Large-scale content distribution: Handling the challenges of securing multimedia content when distributed on a massive scale over the internet.

User privacy concerns: Balancing the need for security with protecting user privacy when implementing security measures.

Cross-platform compatibility: Ensuring multimedia security measures are effective across various devices, platforms, and applications.

Legal and ethical issues: Addressing legal and ethical considerations related to the use, distribution, and protection of multimedia content.

7.2.3 Blockchain: Overview

Blockchain is a ledger technology that provides a distributed and decentralized environment that enables the process of storing records of transactions to be transparent and secured across a network of devices. It was initially designed particularly for cryptocurrencies like Bitcoin as the underlying working technology, but its applications have since expanded to various industries beyond finance. Below is an overview of key concepts and characteristics of blockchain:

i. **Decentralization**
Blockchain is based on a decentralized network of devices (computers), as opposed to traditional centralized systems that are controlled by a single organization. Because every node in the network has a copy of the whole blockchain, it is immune to censorship and single points of failure.

ii. **Distributed ledger**
In basic terms, a blockchain is a distributed ledger, an ever-expanding accumulation of data, or blocks, connected by a chain. An immutable

and secure record of every transaction is created by each block, which consists of a collection of transactions, a reference to the block preceding it, along with a timestamp.

iii. **Consensus mechanism**

Blockchain employs a method of consensus to get agreement on the ledger's state throughout each node. Instances of consensus methods are Proof of Work (found in Bitcoin), Delegated Proof of Stake, Proof of Stake, and many. Consensus makes sure that every node is running an identical edition of the blockchain.

iv. **Immutability**

A block is virtually impossible to tamper with once it is added to the blockchain because it cannot be changed or removed. Cryptographic hash functions and network consensus enable this immutability.

v. **Cryptographic hash functions**

Blockchain secures data inside blocks using cryptographic hash algorithms. A fixed-size hash value that uniquely identifies the block's contents is produced by these functions. A substantial difference in the hash value is produced even with a little alteration in the input data, guaranteeing the reliability of the data.

vi. **Smart contracts**

Self-executing contracts, or smart contracts, have specific conditions encoded directly into the code. When specific requirements are satisfied, they automatically carry out and enforce the terms of a contract. Smart contracts extend the use cases beyond basic transactions by operating on blockchain-based platforms such as Ethereum.

vii. **Public and private blockchains**

Public blockchains are managed by a distributed network of nodes and are accessible to everyone. Anyone may take part in the consensus process, validate transactions, and add to it. Private blockchains, on the other hand, offer more control over permissions and access rights since they are limited to a certain set of participants, sometimes inside an individual organization.

viii. **Permissioned and permissionless systems**

Blockchains that are open or permissionless, such as Bitcoin, permit everyone join without needing authorization. Blockchains with permissions, which are popular in business environments, have limited access and often desire known and verified participants.

ix. **Use cases beyond cryptocurrency**

Numerous sectors, including voting systems, healthcare, banking, logistics, and supply chain management, have found use for blockchain technology. It eliminates the need for middlemen by offering a transparent and safe method of recording and verifying transactions.

x. **Challenges**

Although blockchain technology has many benefits, it also has drawbacks, including issues with interoperability, scalability, energy consumption (in Proof of Work systems), and regulatory concerns. The goal of ongoing study and research is to improve blockchain technology and overcome these problems. The decentralized and secure characteristics of blockchain hold the potential to transform the way data and transactions are managed throughout industries, offering a transparent and trustless platform for a range of applications.

7.3 RELATED WORKS

Intellectual property rights have been preserved by blockchain technology, as industry and academics have begun to examine recently. According to the research currently available, blockchain is believed to be a transparent and trustworthy ledger that can be used to address a variety of copyright protection issues that content creators and owners face, including authenticity, data integrity, rights attribution certificates, transparency, and piracy tracing. The following part of this section provides an outline of the key characteristics and implementation specifics of the current blockchain-based content protection solutions. Using smart contracts, Alka Vishwa (2018) presents a blockchain-based system that ensures multimedia products comply with copyright laws. The transaction details of all the data contributed to the blockchain are stored in the proposed system's data lake, an off-chain centralized storage solution. To guarantee the privacy and validity of the data, it is digitally signed and encrypted on the data lake. Only authorized users who had access permissions along with their signatures in digital form verified by the majority of nodes can access the stored data. Despite guaranteeing user control and privacy over data, the decentralized data management system is only a proof-of-concept that hasn't been tested or put into use in a real environment.

An Ethereum-based digital copyright management system is proposed by Wang Peng (2019) that allows customers and content owners to transact directly eliminating the requirement for a centralized authority. IPFS, a perceptual hash function, digital watermarking, the ElGamal cryptosystem, and smart contracts are employed in the presented system. However, because ElGamal encryption is used to encrypt the entire multimedia material, the approach has a large overhead in terms of CPU time and memory. A safe and dependable blockchain-based real-time eBook market system is presented by Chi (2020), enabling users to publish their content and get paid directly by readers without the need for intermediaries. Blockchain is used by the proposed trading platform to securely handle direct payments and preserve the copyright of purchased content. It provides the following features: confidentiality and data protection; authorization for reading the purchased eBook; verification of a valid buyer; verifiability and non-enforceability of the contents of eBook along with the transactions of direct payment; and safeguards against piracy and illegal distribution of eBook. A book repository is maintained which holds both the book key and the published elliptic curve cryptography-encrypted eBook contents.

A decentralized blockchain-based high-definition video copyright management system was presented by Kishigami (2015). The constructed scheme using this paper's mechanism PoW allows the owners of the copyright themselves to control all activities in it. Other features of the system include encrypting and decrypting the head of an ultrahigh-resolution video (such as 4K and 8K) to balance cryptographic overheads. There is no mechanism, however, in place for providing computational power mining incentives. Still, matters such as access policy management for a media file and cross-platform rendering are missing from this system. Zhao (2018b) introduced the BMCProtector, a system based on the Ethereum blockchain used for music copyright protection that employs AES encryption to protect this asset: protected by smart contracts to pay royalties, music DRM for access regulation, and vector quantization to enforce ownership. While it is very good at copyrighting music files, there are inefficiencies in dealing with other literary works or artistic creations. For example, it cannot protect bootleg recordings of concerts.

Zhaofeng (2018) proposed a DRM to build a blockchain platform and offer a high credit level to the provider of the content along with more security, thereby ensuring the same for the users and the service provider. The platform serves as the storage of information related to content rights to be kept in the blockchain which is resistant to all types of tampers to protect the stealing of information or the copyright from being stolen or misused. Consumers can pay for content consumption with the blockchain-based digital currency, and the coin can be furnished as a multi-signature cryptocurrency digital rights coin on multichain. As a result, developers need a reliable method to guarantee against the printing and copying of redundant content. Meantime, session data encryption and dynamic key agreement were employed to ensure both secure connections and data transmission.

Bhowmik (2017) proposes a new tamper-proofing blockchain framework of multimedia smog, the solution incorporates digital watermarks. This will make distributed images more secure, thereby ensuring integrity. A compressed Sen Sizing (CS)-based self-embedded watermarking algorithms is used to propose the blockchain model from the watermark mater IAL perspective, composed of cryptographic hash and image analog. A cryptographic hash represents multimedia content history pertaining to transaction retrieval. The multimedia blockchain is used for storing the base's value record in textual form stored in multimedia storage elements (MSE). The fields of an image hash are reserved for reference to original multimedia content. In Zhao (2018a), a data concealment technique for video security that is based on blockchains was proposed. Key private data integrity verification improves data concealment. The proposed method comprises three types of data protection: (1) on-chain data protection, which uses a blockchain to archive the signature of the video content and focuses on purity check and security; (2) off-chain data protection, which is use for data hiding algorithm to balance heftiness, visual bias, and embedding capacity; and (3) "data protection management agreement," which uses a smart commitment with registration, inquiry, and transfer contract models.

Li (2021) proposed a chaotic image encryption system related to fingerprints, based on blockchain, that adds protection against security attacks in addition to authentication and traceability. Instead of recording multiple fingerprints with the same length of data, and for the sake of traceability, the distributor's fingerprint is added to the encrypted video and contention matrix. The fingerprints are then compiled with Tardos's collusion-resistant code to be used as one of information. A chaotic map algorithm and reversible watermarking technique are applied to embed the sender's signature or each system distributor's fingerprint into the original picture before any distribution occurs for this content. Qureshi (2019) proposed a content distribution system based on a P2P blockchain. The system employs IPFS networks as storage for multimedia content, Smart Contracts built on Ethereum for execution of atomic payments and proof-of-delivery, perceptual hash functions to ensure content integrity, and homomorphic and symmetric encryption schemes for secure privacy in data transmission. It is worth noting that while this suggested system tries to deal with some of the security problems and privacy threats surrounding an anonymous fingerprinting protocol in distributed environments, it is still just a proof-of-concept that as yet has not been tested or put into any practical use (Deb et al. 2023, Asmitha et al. 2024).

7.4 BLOCKCHAIN INTEGRATION IN MULTIMEDIA SECURITY

This section provides an overview of blockchain technology, explaining its fundamental principles, such as decentralization, consensus mechanisms, and immutability. Emphasis is placed on how these features can be leveraged to address security concerns in multimedia content. This section delves into the ways blockchain technology can enhance multimedia security.

7.4.1 Problem statement

Digital image creation provides an amazing opportunity for photographers and artists to spread their work. But it also brings with it a flood of theft. It is just common practice for images to be copied, shared, and used for commercial purposes without the proper authorization or any recompense to their true authors, adding insult to injury. What's more, without a chain of ownership that is clearly established in writing from start to finish, photographers often find it difficult to establish their own authorship. And then they are unable to enforce their rights (Sahu et al. 2023, Kamil et al. 2023, Deb et al. 2023, Raghunandan et al. 2022).

7.4.2 A solution based on blockchain

The features offered for solving the mentioned challenges with the integration of blockchain include (a) content integrity, (b) digital watermarking, (c) decentralized storage, and (d) smart contracts, as given in Figure 7.1.

a. *Content integrity:* Blockchain's decentralized ledger ensures that multimedia content remains unaltered by providing a tamper-proof record of transactions.
b. *Digital watermarking:* Utilizing blockchain for embedding and verifying digital watermarks enhances the authentication of multimedia content, preventing unauthorized distribution.
c. *Decentralized storage:* Storing multimedia content on a decentralized blockchain network reduces the risk of single points of failure and enhances data availability.
d. *Smart contracts:* Automation through smart contracts facilitates secure and transparent transactions, such as royalty payments for multimedia creators.

Hence, it is evident that blockchain provides certain characteristic features to overcome the challenges of multimedia security. The mapping of each multimedia security challenge along with the features offered by blockchain to address the same is shown in Table 7.1.

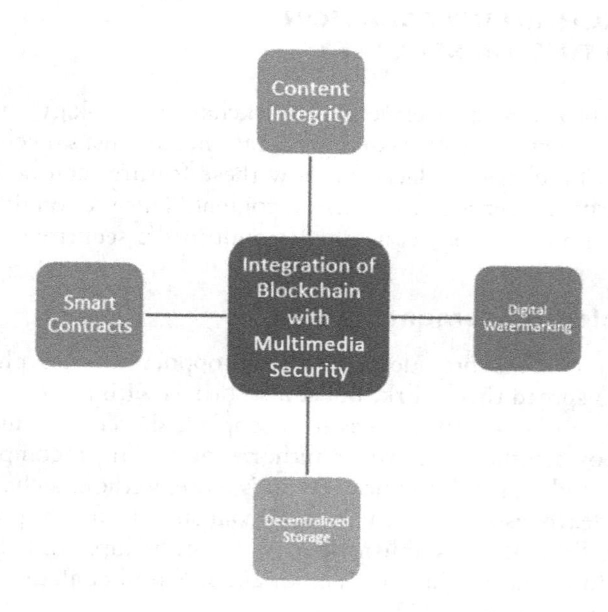

Figure 7.1 Integration of blockchain with multimedia security.

Table 7.1 Mapping of multimedia security issue with the provided features of blockchain

Multimedia security issue	*Blockchain feature as a solution*
Tampering of content	Content integrity
Unauthorized distribution of content	Digital watermarking
Single point of failure and data unavailability	Decentralized storage
Royalty payment transactions	Smart contracts

Further, to solve these problems, a blockchain-based platform builds for picture registration, watermarking, and copyrighting. This platform is based on the blockchain's great features and so offers copyright protection to the creators. Features include:

Secure registration: When pictures are uploaded, they will generate unique digital fingerprints (hashes); these will be lodged on the blockchain. This provides an unalterable record with a fixed date and time showing that the claim to ownership was in fact registered by the photographer himself.

Watermarking techniques: The platform can embed both visible and invisible watermarks invisibly. The visible ones are a strong deterrent; if an image is edited, the hidden watermarks still allow copyright check.

Storage for images: On a mosaic structure for storing images with no single point of inflection, the platform organizes IPFS network.

7.4.3 Benefits

(1) The unalterable timestamps and records captured on blockchain offer artists opening, verifiable evidence of when they created their work. (2) Removing the need for trust and speeding because members of the network fulfill each other's functions using smart contracts. (3) A watermarked image, meanwhile, or a record written into the blockchain provides infringers with a pause-point not deterrent. (4) Potential purchasers can examine ownership records and know that the image they are seeing online is genuinely authorized, which minimizes risk. (5) Basically, potential buyers can also easily check the authenticity and ownership records of any image, reducing the risk of inadvertently using unlicensed material.

7.5 CONCLUSION

Multimedia security involves a multifaceted approach, incorporating various technologies and strategies to protect digital multimedia content throughout its life cycle. As the digital landscape continues to evolve, the field of multimedia security will remain dynamic, requiring continuous innovation to address emerging threats and challenges. The chapter concludes by exploring potential future developments and advancements in the intersection of blockchain and multimedia security. Areas for further research and innovation are identified, paving the way for a more secure and resilient multimedia landscape. While blockchain presents promising solutions for multimedia security, it is important to acknowledge challenges such as scalability, interoperability, and the evolving regulatory landscape. Integration with existing multimedia systems and industry-wide adoption will be crucial for realizing the full potential of blockchain in enhancing the security of multimedia content. Ongoing research and development in this field will likely contribute to overcoming current challenges and expanding the applicability of blockchain technology in multimedia security. In conclusion, blockchain technology emerges as a powerful and promising solution to enhance multimedia security. As the digital landscape continues to evolve, leveraging the decentralized and tamper-resistant features of blockchain can play a pivotal role in safeguarding multimedia content and fostering trust in the digital ecosystem.

REFERENCES

Alka Vishwa, F. K. (2018). A blockchain based approach for multimedia privacy protection and provenance. *IEEE Symposium Series on Computational Intelligence (SSCI)*, 1941–1945.

Asmitha, P., Rupa, C., Nikitha, S., Hemalatha, J., & Sahu, A. K. (2024). Improved multiview biometric object detection for anti spoofing frauds. Multimedia Tools and Applications, 1–17. https://doi.org/10.1007/s11042-024-18458-8

Bhowmik, D. and Feng, T. (2017). The multimedia blockchain: A distributed and tamper-proof media transaction framework. 2017 22nd International Conference on Digital Signal Processing (DSP), London, UK, pp. 1–5.

Chi, J. A. (2020). Secure and reliable blockchain-based eBook transaction system for self-published eBook trading. *PLoS One*, 15, 1–33.

Deb, S., Das, A., & Kar, N. (2023). An Applied Image Cryptosystem on Moore's Automaton Operating on δ(qk)/𝔽2. ACM Transactions on Multimedia Computing, Communications and Applications, 20(2), 1–20.

Deb, S., Pal, S., & Bhuyan, B. (2022). NMRMG: Nonlinear multiple-recursive matrix generator design approaches and its randomness analysis. Wireless Personal Communications, 125(1), 577–597.

Giri, K. J., Parah, S. A., Bashir, R., & Muhammad, K. (2021). Multimedia Security. *Algorithm Development, Analysis and Applications*; Springer: Singapore.

Gurunathan, K., & Rajagopalan, S. P. (2020). A stegano-visual cryptography technique for multimedia security. *Multimedia Tools and Applications*, 79(5), 3893–3911.

Kamil Khudhair, S., Sahu, M., KR, R., & Sahu, A. K. (2023). Secure reversible data hiding using block-wise histogram shifting. Electronics, 12(5), 1222.

Kishigami, J. A. (2015). The blockchain-based digital content distribution system. *2015 IEEE Fifth International Conference on Big Data and Cloud Computing*, 187–190.

Li, R. (2021). Fingerprint-related chaotic image encryption scheme based on blockchain framework. *Multimedia Tools and Applications*, 30583–30603.

Nie, H., Jiang, X., Tang, W., Zhang, S., & Dou, W. (2020). Data security over wireless transmission for enterprise multimedia security with fountain codes. *Multimedia Tools and Applications*, 79, 10781–10803.

Qureshi, A. a. (2019). Blockchain-based P2P multimedia content distribution using collusion-resistant fingerprinting. *Proceedings of APSIPA Annual Summit and Conference 2019*, 1606–1615.

Raghunandan, K. R., Dodmane, R., Bhavya, K., Rao, N. K., & Sahu, A. K. (2022). Chaotic-map based encryption for 3D point and 3D mesh fog data in edge computing. IEEE Access, 11, 3545–3554.

Sahu, A. K., Umachandran, K., Biradar, V. D., Comfort, O., Sri Vigna Hema, V., Odimegwu, F., & Saifullah, M. A. (2023). A study on content tampering in multimedia watermarking. SN Computer Science, 4(3), 222.

Wang Peng, L. Y. (2019). Secure and traceable copyright management system based on blockchain. *Distributed Computing: Science Method*, 1243–1247.

Zhao, H. a. (2018a). A blockchain-based data hiding method for data protection in digital video: First International Conference, SmartBlock 2018, Tokyo, Japan, December 10–12, 2018, Proceedings. In *International Conference on Smart Blockchain* (pp. 99–110).

Zhao, S. A. (2018b). BMCProtector: A blockchain and smart contract based application for music copyright protection. *ICBTA 2018: Proceedings of the 2018 International Conference on Blockchain Technology and Application*, 1–5.

Chapter 8

Cybersecurity-based artificial intelligence healthcare management system

Panem Charanarur, Srinivasa Rao Gundu, and Monalisa Sahu

8.1 INTRODUCTION

When it comes to cybersecurity, the healthcare business is among the most vulnerable among industrial sectors. In this age of COVID-19, the healthcare sector all around the globe is becoming more and more vulnerable to cyber threats. Several organizations that specialize in cybersecurity have observed an alarming increase in the number of cyberattacks that have occurred since the beginning of the COVID-19 outbreak. Cybercriminals have considered nursing homes and other components of the healthcare system to be a primary target for a considerable amount of time [1]. As a result of recent assaults on several big healthcare systems and hospitals, the security vulnerabilities of the majority of respected healthcare firms have been brought to individuals' attention. Within the context of the battle against the COVID-19 pandemic, the healthcare industry is at the forefront of the struggle [1].

It is thus reasonable to assume that this vital sector is protected from cybercriminals; nevertheless, this is not the case. In the period after COVID-19, there has been a steep spike in the number of cyberattacks, with the healthcare industry serving as the target of most of these attacks. Because of the sensitive category of patient health–related data, the healthcare-based industry places a high priority on ensuring that patients' privacy and security are fully protected. For a few decades, technology usage has increased by healthcare organizations. This includes utilizing cloud storage, new medical equipment, artificial intelligence (AI), machine learning techniques for patient and health profile protection, and many other technologies [2].

Currently, the healthcare business is a result of these technological improvements, which have significantly simplified the work that providers are responsible for. But the likelihood of cyberattacks is higher than it has ever been before. The databases of the majority of hospitals are not effectively safeguarded, and medical professionals, such as physicians and nurses, are not always aware of the cyber threats that might potentially influence their practice [3]. It may take many weeks to notice a cyberattack because of the sensitive nature of health information technology. Despite being unaware of

 DOI: 10.1201/9781032663647-8

the invasion, healthcare personnel continue to use a system that has been corrupted. Digital health has already demonstrated that it can enhance our own and our family's health while also helping us to be more productive. Digital health increases access to healthcare information and services, improves delivery quality, and allows for a far more personalized application of healthcare to patients, lowering healthcare delivery inefficiencies. Simply put, digital health entails linking the systems, tools, digital medical devices, and services that provide crucial healthcare to each of us, as well as providing previously inaccessible data insights to each stakeholder in the healthcare delivery ecosystem [4].

Manufacturers, hospitals, and institutions must work together to handle cybersecurity concerns, which cannot be completely avoided. It's crucial to strike a balance between patient safety advancing new technologies and improving device performance. The healthcare industry's overall complexity adds to the necessity of finding a solution to this problem. Healthcare organizations, particularly hospitals, must find out how to protect legacy systems and devices that support all aspects of their operations while also introducing new connected devices to their networks that were not designed with security in mind, to begin with. Many companies that develop and manufacture class 3 medical devices are unaware of even the most basic security measures that must be followed for the devices to be made properly. However, it also allows for suitable access and authorization controls for both the device and the data it produces. By distributing these devices throughout their networks, healthcare organizations and hospitals are introducing new dangers and vulnerabilities, which could have a substantial impact on patients' lives. The benefits of digital health are accompanied by a danger that, if not adequately handled and minimized, has the potential to cause significant harm. Fortunately, this danger is completely avoidable.

8.2 AIM AND OBJECTIVES OF THE CHAPTER

The adoption of AI in cybersecurity could be hampered or even lead to significant problems for society if there are no proper processes and policies in place, according to the USNIST, an organization dedicated to advancing AI. Healthcare organizations that transitioned swiftly to digital healthcare initiatives are now confronted with concerns that other industries have been addressing for years. This might result in billions of dollars being spent each year, affecting millions of people. In addition to external cybersecurity concerns, healthcare providers must occasionally deal with internal dangers. Human error or a breach of an employment contract is the source of these internal risks to enterprises. There are three forms of internal attacks, according to various case studies: employee or contractor carelessness or negligence, criminal or malicious insider, and credential theft [5].

8.3 LITERATURE REVIEW

According to the findings, the scarcity of healthcare professionals, which comprises more than 17 million individuals, is making the impacts of an aging workforce even more devastating. The use of supercomputers might make it simpler to discover novel approaches to the treatment of illnesses. It is feasible that advances in AI might reduce the infectious potential of viruses in a single day rather than over many years. Technology has the potential to improve the delivery of healthcare while simultaneously lowering costs. This will be accomplished by simplifying operations, boosting efficiency, and reducing the burden of personnel.

In 2019, Mahoney joined Lin and his colleagues as a member of the group. Many companies are investigating the potential of AI to assist them in improving their problem-solving abilities and boosting their production via improved use of population health technologies. Numerous companies are now developing AI doctors that can directly give patients fundamental medical advice in an attempt to lessen some of the strain that is associated with the treatment of medical conditions that are more complex. According to a chapter that was published not too long ago, using AI in the chapter on the brain would significantly increase our knowledge of the complex systems that it has. When new information about particular illnesses becomes available, AI can make it easier for medical practitioners to keep themselves informed of the latest clinical developments. The use of machine learning algorithms can significantly reduce the symptoms of many mental illnesses if not completely cure them.

It has been stated that recent developments in AI have resulted in significant advancements in thoracic surgery (Other sources mention comparable findings). Robots have made it possible to perform surgical treatments in a more timely and precise manner with more precision. The technology can duplicate the surgeon's hand gestures with pinpoint precision at the same time as the physician is providing the robot with particular instructions. According to the findings, heart disease is one of the leading causes of mortality around the globe. The presence of blood clots in the arteries is associated with an increased risk of cardiovascular disease, especially in younger persons who maintain unhealthy lifestyles. Cardiologists may be able to make better decisions if they use data, AI, and machine learning. Robert was the one who submitted the essay in the year 2019. In the disciplines of health and nursing, the creative ideas that are made feasible by AI are proving to be beneficial. As a consequence of the development of nursing robots, there will be a change in the function that nurses are responsible for. The duties that are required of these individuals include providing assistance with mobility, taking vital signs, administering medication, and adhering to guidelines for contagious diseases. Robots will improve the quality of care that nurses can deliver to patients by freeing up their time.

8.4 PROBLEM STATEMENT

It is essential that there should be some semblance of human monitoring and control over AI systems. An insufficient level of cybersecurity in the protection of open-source models may potentially lead to chances for hacking. If there were restrictions placed on the distribution and sharing of data and codes, it could be possible to get a more accurate picture of the associated security threats.

8.5 OBJECTIVES OF THE CHAPTER

Our business focuses on developing a security system that includes motion detection, photo analysis, and digital owner notification. The Raspberry Pi, a single-board computer, will be responsible for most of the work. Our primary objectives include conducting a chapter, providing necessary connections for a motion detector and camera, updating Raspberry Pi settings for email sending, creating a fully operational surveillance prototype, and developing a technique for real-time object recognition and tracking in video. The Raspberry Pi will be used to create a surveillance system component of our project.

8.6 METHODOLOGY (SECURED CYBER INTELLIGENCE–BASED HEALTHCARE)

Medjacking is the act of attempting to damage a patient by attacking and manipulating a medical device or instrument. Any malfunctioning medical tools at a hospital or clinic can be extremely stressful and potentially dangerous. Any medical instrument's inaccurate diagnostic results could result in an improper prescription. If any medical devices fail to function effectively, they may cause injury to patients, possibly leading to death. Medjacking often targets well-known individuals and smears healthcare organizations. The use of medical devices and instruments can be made safer with the help of AI. To summarize, in the SDTS era, an ICT framework will be used to develop rural concerns, as well as to build a linked, technologically equipped, and long-lasting infrastructure [6].

The smartness of an SDTS era is determined by several factors:

1. A technological infrastructure
2. Environmental efforts
3. Public transit that is both effective and functional
4. Reliable and forward-thinking city/state/location plans
5. People who can live and work in the city/state/location while taking advantage of its resources [7]

8.7 JUSTIFICATION

Software or hardware referred to as "artificial intelligence" may process complex inputs such as pictures, videos, text, music, or sounds and produce the expected result. Some worry that, in the future, robots may be able to make complex medical choices on their own, thanks to AI-driven computers' ability to make judgments with little human involvement. About AI and health, it is critical to separate fact from myth [8]. Commercial–public partnerships are crucial to SDTS' success. Most of the labor needed to build and sustain a data-driven environment is beyond local government's authority. Intelligent surveillance cameras may need cross-industry collaboration and technology. IT at healthcare facilities is sensitive; therefore, it may take weeks to detect a hack. Healthcare professionals are ignorant of breached systems when they use them. This might cost billions and affect millions of people [9]. Data analysts must review SDTS era data to find mistakes and make changes. This is in addition to smart city, state, or place technology. The COVID-19 virus must be detected quickly, sick people quarantined, and contacts found.

To get over accessibility and distance issues, you may use IoT and CPS protocols, together with GPS and Wi-Fi. As the recent COVID-19 pandemic has shown, the IoT may help contain infectious disease outbreaks. As an example, smart thermometers capture data for better analysis and diagnosis by measuring the patient's temperature [10]. Quarantine compliance is another area where the Internet of Things (IoT) might be useful. Keeping tabs on quarantine and local trends is made simple with the use of tracking and tracing technologies, patients, and their cellphones or mobile devices.

Machine-to-machine communications (M2M) is a term that refers to the communication between machines. Mesh Networks are a type of network that consists of a series of inter. To successfully address healthcare cybersecurity problems, a new and more precise layer of trust throughout the digital health landscape is necessary, which can reduce the impact of a security breach in health networks. Simply said, once a system's IP address is discovered, it may be hacked. Healthcare systems, databases, medical equipment, and even everyday operating objects like printers that are connected to a digital health network fall into this category. Cloaking systems and devices have proven a very successful alternative for keeping any unknown or unauthorized people from seeing health systems or equipment, whether legacy or newly new. Companies can now totally prohibit a bad actor from gaining access to essential and sensitive systems or devices, putting patient data at risk or jeopardizing device operation [11]. This functionality is placed in "pre-network session" formation, which adds no additional complexity to the topology of a network and considerably decreases deployment time. Every healthcare business is recognizing the importance of a proactive approach that reduces "time to security" [8]. Several significant changes are on the horizon in the healthcare industry. External factors and technology improvements will continue to grow in importance, posing new threats. Cybersecurity and privacy must now, more than ever, be completely integrated into the development and deployment of innovative healthcare services and solutions by design. Not only to realize the promise of the future of health, but also to ensure a safe and secure future, industry actors must also respond effectively to the forces of change and the challenges ahead [12].

8.8 NEED FOR SDTS ERA

As of now, 54% of the global population resides in urban areas; by 2050, with a projected

There are 2.5 billion more people living in urban areas, which is projected to climb to 66%. As the global population and economy continue to expand, so does the urgency of the need to ensure the long-term viability of these resources. To make smarter and more efficient use of resources in a more urbanized world, residents and city governments may collaborate in the age of SDTS. Improvements to the SDTS era also bring new value to existing infrastructure by allowing for new revenue streams and increased operating efficiency, which saves governments' and individuals' money. SDTS era places a major focus on sustainability as part of its efforts to promote urban efficiency and public well-being. Cities have environmental advantages, such as smaller geographic footprints, but they can also have disadvantages, such as dependency on fossil fuels for energy. On the other side, smart technology may be able to alleviate these negative consequences, for as by installing an electric transportation system to reduce emissions [13].

Because autonomous cars are predicted to lower people's desire to possess a car, the number of automobiles in cities should decrease as a result of such ecologically benign transportation choices. Developing such long-term solutions might be advantageous to both the environment and the economy. Smart cities offer many advantages, but they also have many disadvantages. Government officials who allow a large number of citizens to participate are among them. Individuals must also engage with the business and governmental sectors so that everyone may make a positive contribution to society [14]. The following are the security precautions that must be performed.

1. Leading healthcare organizations will begin to investigate new security technologies that might help them prevent security risks rather than only recognize them after they occur, resulting in a considerable increase in the level of protection provided while lowering operational expenses.
2. Network masking will be possible, thanks to better security technology, allowing only authorized users to observe a system or device. When IT systems are only accessible to known and authorized persons, the risk of a hack is practically eliminated.
3. Cybersecurity and patient safety problems must be addressed by everyone participating in the digital health ecosystem. Preventive modeling

and cloaking technologies can be used "pre-market" to prevent non-malicious insiders from gaining unauthorized access and exploiting security flaws.

Example: Smart technologies are being developed and used in cities all around the world at various levels. There are a few places, though, that are ahead of the curve and leading the way for really smart cities. The following are a few examples:

1. *Availability:* Data must be available in real time and with dependable access to be utilized to monitor various aspects of smart city infrastructure.
2. *Accessibility and accuracy:* The information must be both accessible and accurate. This includes guarding against meddling from the outside.
3. *Confidentiality:* Private and confidential information must be kept private and safeguarded from unwanted access. Firewalls and data anonymization are two examples of this.
4. *Accountability:* Users of sensitive data systems must be held accountable for their activities and interactions.
5. User logs should keep track of who is accessing the data to guarantee responsibility in the event of an issue.

Smart technology will help cities become more efficient and sustainable. Electric cars benefit from solar-to-electric charging, and traffic and crime are monitored by networked cameras. New infrastructural discoveries must be polished before they can be applied in the real world.

8.9 SATELLITE INTERNET ACCESS

In locations where wired Internet connectivity is difficult or unavailable, satellite Internet access is a comparatively inexpensive option (remote areas). Moving hosts can also use satellite connections, such as the Panasonic ex Connect for Internet access and GSM phone service on long-haul flights. A satellite system is often designed for one-way communication (TV, radio). Downlink bandwidth is significantly less expensive than uplink bandwidth [15]. To allow smart cities, data management is a vital component of the IoT implementation. Data collection, processing, and distribution are the most common responsibilities. All parts of data standards, quality, and utilization are referred to as "data collection." When dealing with enormous amounts of data, they are required. Data standards ensure that data collection techniques are standardized across the board. According to a 2015 poll, electronic health records have surpassed chapter-based health records in developed countries by more than 80% in developed areas and up to 40% in emerging nations [16].

The number of digital consumer health services accessible in Canada has more than doubled in only two years, causing the Australian Digital Health Agency to be established in July 2016 to supervise the country's digital health development and delivery. These are just a few instances of how technology is being increasingly used to improve worldwide access to healthcare data. However, with the potential benefits of digital EHR deployment comes the risk of risks, many of which may reveal themselves indirectly, such as breaches at third-party EHR providers. According to IBM Managed Security Services (MSS) figures from 2016, insiders were responsible for 68% of all network assaults against healthcare organizations. The most prevalent attack vector was, predictably, the use of malicious data supplied by bad actors to try to change or disrupt the functioning of target systems. As the black market value of healthcare records packed into comprehensive person profiles grows, attackers will increasingly target the healthcare business [17].

A safe environment is now a need. Now, more than ever, organizations must transform a point product–based collection of security solutions into an integrated security immune system. Healthcare organizations continue to be devastated by insiders, both malicious and unwitting. According to IBM Managed Security Services data, insiders carried out considerable percentage of all network attacks against healthcare firms in 2016, while hostile actors were engaged in more than one-third of those attacks. An incident in April 2016 exemplified the danger presented by someone motivated by malice. Insiders may do just as much damage as external threats if they contribute dangers to the environment unknowingly or inappropriately. An organization's failures and vulnerabilities, such as falling for phishing schemes, misconfiguring servers, or misplacing laptops, may offer cybercriminals a wide-open backdoor into its networks and allow them to carry out attacks. Although stolen computers are a concern in every sector of the economy, the healthcare and pharmaceutical sectors rank second among all enterprises in terms of the reported losses they have sustained as a result. How many instances of laptop theft get unreported? How much sensitive information, as a direct consequence of this, falls into the wrong hands?

Cyber is the driving force behind all these astonishing technological advancements. The future healthcare system is now in the first phases of preparation, but the risks associated with insufficient cybersecurity will inevitably escalate as it materializes. One of the greatest healthcare data breaches ever recorded resulted in the exposure of personal information belonging to around 80 million patients. The breach is believed to have originated from a phishing attack. Transportation, energy, healthcare, water, and waste management are essential urban infrastructures and services that are enhancing in terms of dependability, effectiveness, and availability. The objective of establishing a "smart city" is to tackle several challenges, such as enhancing the socioeconomic situations of urban residents, minimizing energy use, safeguarding the environment, and enhancing the transportation infrastructure of the city. Politicians are increasingly incorporating the notion of "smart cities" into their policy agendas [18].

Approximately 50% of the world's population now resides in metropolitan areas, and the pace of population increase in these regions is rapidly increasing. Transportation, energy, healthcare, water, and waste management are crucial urban infrastructures and services that are enhancing in terms of dependability, effectiveness, and availability. The objective of establishing a "smart city" is to tackle several challenges, such as enhancing the socioeconomic situations of urban residents, minimizing energy use, preserving the environment, and enhancing the transportation infrastructure of the city.

Several governments are increasingly prioritizing the notion of "smart cities" as it gains momentum. An integral aspect of the ever-developing notion of a "smart city" is the seamless facilitation of data transmission among the many linked services. For example, in the event of an emergency, the transport system may create information that enables users to connect with emergency service networks, such as hospitals and government institutions. Data integration among public utilities, such as water, gas, and electricity, may also be advantageous for municipal operations, including taxes. Given the ongoing flood of this data, it is imperative to establish appropriate protocols. The concept of a "smart city" is continuously growing in parallel with the advancements in AI [19]. The primary objective of AI is to facilitate the automation of systems to optimize the efficiency and productivity of urban infrastructure, while concurrently cutting costs, minimizing resource use, and enhancing service quality. AI can tackle several urgent difficulties faced by smart cities today. These include issues such as transportation congestion, cybersecurity, privacy, and governance. Within the realm of smart cities, an extensive network of sensors is used to provide the required services. Hence, the infrastructure of a smart city should include networked devices such as sensors and actuators that are trustworthy and reliable. These devices are responsible for gathering, analyzing, and transmitting data to ensure the delivery of dependable municipal services. Nevertheless, the adoption of smart cities presents the issue of managing several networks, each with distinct cybersecurity demands. Service-authorized communication may occur on low-end, low-powered devices with little storage space, as is often the case. These devices may not have important security features such as cryptographic capability or authentication procedures due to limitations in CPU and storage capacity.

As a result, the apps that provide smart city services are susceptible to possible assaults. Consequently, there is a substantial likelihood that the cyber realm of a smart city would be susceptible to major security breaches. In the absence of security measures, a cyber-physical system is susceptible to unauthorized access by prospective attackers. Every potential weakness in the system might be taken advantage of. In the context of smart cities, the interconnection of all devices, independent of their functionalities, facilitates the vulnerability of lower-end equipment to potential attackers, therefore granting them access to the service system. The main emphasis of cybersecurity is ensuring the safety of connected and Internet-based services. This is because several elements of smart cities are interconnected with the Internet. Moreover, it expands upon the concept of safeguarding not just the transmission of data via networks, but also the physical components and programmers of a computer system. The result is a system that exhibits both stability and dependability. Implementing cybersecurity measures in live monitoring enhances the efficacy of safeguarding against external intrusions. The transition of interconnected cities into intelligent cities has led to a substantial influx of data being exchanged among the several organizations engaged in this domain [20]. The emergence of "smart cities" is a direct result of the transformation of "linked cities."

Cities are increasingly adopting digital integration to foster sustainable economies and ecosystems. The interactions between urban residents and their surroundings produce vast quantities of data. Smart cities must ensure the provision of adequate support for these excess services and resources. To satisfy the needs of a large population, cities must enhance their existing networks by using AI and IoT technologies. This cutting-edge technology can effectively manage resources, assets, and supply chains. Performing data analysis, implementing reasoning algorithms, and managing sensor networks all are challenging endeavors. Your well-being may be at risk if you depend on systems and technology that are susceptible to cyberattacks. Selecting an appropriate risk mitigation strategy to address potential concerns is a complex process. It is well acknowledged that IoT devices include inherent security vulnerabilities. Manufacturers are not legally required to comply with any safety regulations or guidelines. As a result, they lack knowledge about how security vulnerabilities might potentially expand the vulnerability of a system to attacks.

The IoT is a collection of interconnected computing devices that provide remote monitoring and control capabilities for a city's service infrastructure. This kind of cyber-physical infrastructure may be equipped with technological smarts and automation capabilities via the application of AI [21, 22]. Network heterogeneity poses the greatest risk to the successful application of smart city technology. Cyber threats can bring about both failure and disaster in an organization. The term "cyber risks" may refer to a broad variety of situations. As a consequence of this, in addition to the conventional security precautions, AI that is based on machine learning has been used to ensure the safety of cyber-physical systems. According to the ABI Chapter, 44% of the money spent on cybersecurity for smart city efforts would go toward the management of infrastructure, including electricity, healthcare, public safety, transportation, water, and waste treatment [23–25].

8.10 FINDINGS

The recent developments in areas such as machine learning and ubiquitous computing are helping to accelerate the process of merging smart city technologies with healthcare technology. As a direct result of the greater sharing of sensitive health data about people, there has been a rise in the amount of concern around potential intrusions of privacy. Healthcare systems that make use of AI may be able to create applications that are not just efficient financially but also centered on the needs of patients.

8.11 CONCLUSIONS AND FUTURE PERSPECTIVES

The machine learning systems must be able to give the doctor the information he or she needs from the account. Dr. Richard Branson says that having a reason for a choice gives it more weight and makes it easier to understand. The new age is the "Self-Driven Technology Set". AI-based society not only makes people's lives better, but it also helps solve a lot of problems. The elite society would be built on the chapter of facts. Data science is a branch of science, numbers, and technology that deals with data and everything to do with it. With the help of AI, health problems could be found more correctly. This is where smart technologies and the human-in-the-loop system could help by spotting the illness and shortening the time it takes for a person to get care. When health problems are taken care of quickly, they can make the situation less bad overall. COVID-19 has also made a lot of healthcare systems too busy, which has made it easier for bad people to target them and find cybersecurity flaws. AI might be able to help fix problems with Internet security so that a safe human-in-the-loop solution to mental health problems can be made. With AI-based IoT-connected cloud-oriented robotic process atomization, society would work better. If these technologies were always set up the same way, they would work well together, with one coming before the other. These technologies are not the same, and there may be a set of technologies in which neither one comes before the other. This chapter will give hints about the design and methods of the next age of technology, as well as its needs and gaps. Leading healthcare firms will begin to look at new security solutions that can prevent security risks, rather than merely identifying breaches after they have happened, and this will greatly increase the degree of protection given while also reducing operating expenses. Network cloaking will be possible, thanks to new security technologies, enabling only authorized users to view a system or device. When IT systems are only visible to known and authorized identities, the possibility of a hack is almost eliminated. Everyone in the digital health ecosystem must be accountable for cybersecurity and resolving patient safety concerns. Preventive models and cloaking technologies can be used "pre-market" to defend against nonmalicious insiders gaining unauthorized access and introducing vulnerabilities.

REFERENCES

1. H. F. Al-Turkistani and H. Ali, "Enhancing users' wireless network cyber security and privacy concerns during COVID-19," 2021 1st International Conference on Artificial Intelligence and Data Analytics (CAIDA), 2021, pp. 284–285. doi: 10.1109/CAIDA51941.2021.9425085

2. J. Ahmed and Q. Tushar, "Covid-19 pandemic: A new era of cyber security threat and holistic approach to overcome," 2020 IEEE Asia-Pacific Conference on Computer Science and Data Engineering (CSDE), 2020, pp. 1–5. doi: 10.1109/CSDE50874.2020.9411533
3. A. A. Shammari, R. R. Maiti, and B. Hammer, "Organizational security policy and management during Covid-19," SoutheastCon 2021, 2021, pp. 1–4. doi: 10.1109/SoutheastCon45413.2021.9401907
4. E. B. Sloane, V. Gehlot, N. Wickramasinghe, and R. Silva, "Using community care coordination networks to minimize hospitalization of COVID-19 patients," SoutheastCon 2021, 2021, pp. 1–4. doi: 10.1109/SoutheastCon45413.2021.9401927
5. Asmitha, P., Rupa, C., Nikitha, S., Hemalatha, J., and Sahu, A.K., Improved multiview biometric object detection for anti spoofing frauds. Multimedia Tools and Applications, pp. 1–17 (2024). https://doi.org/10.1007/s11042-024-18458-8
6. H. Cai, T. Yun, J. Hester, and K. K. Venkatasubramanian, "Deploying data-driven security solutions on resource-constrained wearable IoT systems," 2017 IEEE 37th International Conference on Distributed Computing Systems Workshops (ICDCSW), 2017, pp. 199–204. doi: 10.1109/ICDCSW.2017.15
7. A. K. Siddhu, A. Kumar, and S. Kundu, "Review chapter for detection of COVID-19 from medical images and/or symptoms of patient using machine learning approaches," 2020 9th International Conference System Modeling and Advancement in Chapter Trends (SMART), 2020, pp. 39–44. doi: 10.1109/SMART50582.2020.9336799
8. G. Guannna, "British community health care system based on big data and artificial intelligence," 2020 International Conference on Robots & Intelligent System (ICRIS), 2020, pp. 248–252. doi: 10.1109/ICRIS52159.2020.00069
9. M. Ma, M. Skubic, K. Ai, and J. Hubbard, "Angel-Echo: A personalized health care application," 2017 IEEE/ACM International Conference on Connected Health: Applications, Systems and Engineering Technologies (CHASE), 2017, pp. 258–259. doi: 10.1109/CHASE.2017.91
10. O. Duda, V. Pasichnyk, N. Kunanets, R. Antonii, and O. Matsiuk, "Multidimensional representation of COVID-19 data using OLAP information technology," 2020 IEEE 15th International Conference on Computer Sciences and Information Technologies (CSIT), 2020, pp. 277–280. doi: 10.1109/CSIT49958.2020.9321889
11. J. J. E. Macrohon, and J.-H. Jeng, "A real-time COVID-19 data visualization and information repository in the Philippines," 2021 9th International Conference on Information and Education Technology (ICIET), 2021, pp. 443–447. doi: 10.1109/ICIET51873.2021.9419591
12. Y. Chen, C. K. Leung, S. Shang, and Q. Wen, "temporal data analytics on COVID-19 data with ubiquitous computing," 2020 IEEE International Conference on Parallel & Distributed Processing with Applications, Big Data & Cloud Computing, Sustainable Computing & Communications, Social Computing & Networking (ISPA/BDCloud/SocialCom/SustainCom), 2020, pp. 958–965. doi: 10.1109/ISPA-BDCloud-SocialCom-SustainCom51426.2020.00146
13. A. Rajkumar et al., "Visualizing effects of COVID-19 social isolation with residential activity big data sensor data," 2020 IEEE International Conference on Big Data (Big Data), 2020, pp. 3811–3819. doi: 10.1109/BigData50022.2020.9377830.
14. R. K. Ramesh, R. Dodmane, S. Shetty, G. Aithal, M. Sahu, and A. K. Sahu, A novel and secure fake-modulus based Rabin-3 cryptosystem. Cryptography, 7(3), p. 44 (2023).

15. N. Wang and B. Mao, "The chapter on the problems of smart old-age care in the background of smart city construction," 2019 International Conference on Intelligent Transportation, Big Data & Smart City (ICITBS), 2019, pp. 151–154. doi: 10.1109/ICITBS.2019.00043
16. K. Axel and I. S. Khayal, "Modeling 'Thriving Communities' using a systems architecture to improve smart cities technology approaches," 2018 IEEE International Smart Cities Conference (ISC2), 2018, pp. 1–2. doi: 10.1109/ISC2.2018.8656893
17. G. K. Garge, C. Balakrishna, and S. K. Datta, Consumer health care: Current trends in consumer health monitoring. IEEE Consumer Electronics Magazine, 7(1), pp. 38–46 (2018). doi: 10.1109/MCE.2017.2743238
18. R. Lee, C. Lai, and S. Chiang, "Design of health-care home gateway (HCHG) over digital cable television network," 2007 Digest of Technical Chapters International Conference on Consumer Electronics, 2007, pp. 1–2. doi: 10.1109/ICCE.2007.341319
19. P. Patro, K. Kumar, G. S. Kumar, and A. K. Sahu, Intelligent data classification using optimized fuzzy neural network and improved cuckoo search optimization. Iranian Journal of Fuzzy Systems, 20(6), pp. 155–169 (2023).
20. F. Young, L. Zhang, R. Jiang, H. Liu, and C. Wall, "A deep learning based wearable healthcare IoT device for ai-enabled hearing assistance automation," 2020 International Conference on Machine Learning and Cybernetics (ICMLC), 2020, pp. 235–240. doi: 10.1109/ICMLC51923.2020.9469537
21. R. S. Rao, L. R. Kalabarige, B. Alankar, and A. K. Sahu, Multimodal imputation-based stacked ensemble for prediction and classification of air quality index in Indian cities. Computers and Electrical Engineering, 114, p. 109098 (2024).
22. S. Deb, A. Das, and N. Kar, An applied image cryptosystem on Moore's automaton operating on $\delta(qk)/\mathbb{F}2$. ACM Transactions on Multimedia Computing, Communications and Applications, 20(2), pp. 1–20 (2023).
23. K. S. Roy, S. Deb, and H. K. Kalita. A novel hybrid authentication protocol utilizing lattice-based cryptography for IoT devices in fog networks. Digital Communications and Networks (2022). https://doi.org/10.1016/j.dcan.2022.12.003.
24. K. R. Raghunandan, R. Dodmane, K. Bhavya, N. K. Rao, and A. K. Sahu, Chaotic-map based encryption for 3D point and 3D mesh fog data in edge computing. IEEE Access, 11, pp. 3545–3554 (2022).
25. S. Deb, S. Pal, and B. Bhuyan, NMRMG: Nonlinear multiple-recursive matrix generator design approaches and its randomness analysis. Wireless Personal Communications, 125(1), pp. 577–597 (2022).

Chapter 9

A taxonomical review on cloud security and its solutions

Panem Charanarur and Srinivasa Rao Gundu

9.1 INTRODUCTION

Models for offering cloud computing services include the ones listed below as examples. When it comes to providing cloud services, there are three fundamental models to consider, each of which is becoming more established and common with each passing generation. For this, there are many various approaches to consider, including software as a service (SaaS), platform as a service (PaaS), and infrastructure as a service (IaaS), to name but a few. A few of these strategies include software development, platform and infrastructure as a service, and cloud computing, among others [1–3]. To access programs hosted on service provider infrastructure, users must connect to them online. This is called SaaS or cloud computing, depending on whom you ask. These strategies assist the customers of software offered under the SaaS business model, who are typically end users who make subscriptions to readily available programs. The SaaS model has also been associated with a pay-and-use feature allowing end users to access the software through a web browser without dealing with the headaches of installation, maintenance, or making a significant upfront payment. Some popular SaaS apps include SalesForce, Google Apps, and Google Docs [4–6].

User awareness is an essential component of SaaS security from a security viewpoint. However, the SaaS provider must hold to a set of security conditions to ensure users adhere to the critical security protocols while using the service. Examples of these requirements include multi-factor authentication, complicated passwords, and password retention. An additional component that SaaS providers should have in place is the adoption of security measures to secure customers' data and to guarantee that it is available for permitted usage at all times. In computing, the phrase Platform-as-a-Service refers to a collection of software and development tools stored on a service provider's servers and available from any location on the Internet. It provides developers with a platform to construct their apps without worrying about the underlying mechanics of the service they rely on for support. It also makes it easier to manage the software development life cycle, from planning to maintenance, efficiently and effectively, thanks to the PaaS architecture [7–10].

DOI: 10.1201/9781032663647-9

The platform also uses programming languages such as VC++, Python, Java, etc., to allow users to construct their apps on top of it. Many developers and programmers now depend on PaaS firms such as WordPress, Go Daddy, and Amazon Web Services to build websites and host online applications. According to the PaaS paradigm, security is a shared responsibility that developers and service providers must handle in equal measure. Example: When developing applications, developers must follow security standards and best practices to guarantee that the applications are safe and secure. A programmer, for instance, must certify that the software is free of flaws and vulnerabilities before exposing it to the general public. Aspects of this process that are equally important include detecting and correcting any security flaws that attackers may exploit to get access to and compromise users' data. For developers, the dependability of PaaS technology, on the other hand, is critical to providing a safe and secure environment for application development. For example, several programming environments, such as C++, are well-known for having poor memory management, enabling attackers to conduct assaults against their victims, including stack overflows [3]. A lack of sufficient authentication in some relational database management systems (RDBMSs), such as Oracle, may also be exploited by attackers. Oracle, for example, allows users who have been granted admin permissions at the operating system level to access the database without needing a username and password [11–15].

A cloud computing paradigm in which a cloud computing service provider is dedicated to the resources that are only shared with contractual customers that pay a per-and-use charge to the cloud computing service provider is called IaaS. In particular, one of the critical benefits of the Equipment as a Service model is that it removes the need for a significant initial investment in computer infrastructure such as networking devices, computer processors and storage capacity, and servers. The technology may also be used to quickly and cost-effectively increase or reduce the amount of computer resources available to a user. Today, with the proliferation of cloud delivery systems, it may take time to determine the boundaries of one's security responsibilities. Security is the responsibility of cloud service providers (CSPs) and clients using their services. As seen in Illustration 5, the duties of cloud computing service delivery models are outlined. Cloud computing services include infrastructure (IaaS) offerings such as Amazon Web Services, Cisco Meta-cloud, Microsoft Azure, and Google Compute Engine (GCE). It is important to note that customer-facing infrastructure is critical regarding security since it acts as the first line of defense for the system's perimeter [16–20].

In this environment, attackers may use various strategies to target the infrastructure, including denial of service (DoS) attacks and malware distribution campaigns. Most of the time, the security of a PaaS solution is the service provider's responsibility. Cloud Models and Architectures: An Introduction Determining the Kind of Cloud an Institution Should Use is the first and most crucial stage in cloud deployment, as this will allow for a smoother installation process. The second and final step is deployment during the cloud deployment process. According to the authors, institutions that have failed to execute a deployment plan have done so as a result of selecting the incorrect kind of cloud infrastructure. Organizations must assess their data before deciding on the cloud infrastructure to prevent failure. While many consumers consider security when signing up for cloud services, many do not because they have a misconception of the efficiency of the protection given by cloud services in and of itself. When it comes to keeping their data secure, many businesses that use cloud computing depend only on the security measures employed by CSPs. This may allow hostile actors to exploit client-side vulnerabilities to attack the systems of one or more tenants due to the situation [21,22].

To mention a few examples, public cloud, private cloud, community cloud, and hybrid cloud all are concepts that are being explored. In some areas, the public cloud is often called an external one, as with the Amazon Web Services (AWS) public cloud. This cloud is accessible to all users or large groups through the Internet, with CSPs retaining control over the environment. Customers may access any data made accessible on the network using this service, which the service provider manages. A cost-effective and scalable means of implementing information technology solutions is made feasible via public cloud computing. Because of the Internet connection, various security dangers are introduced into the system, including DoS attacks, malware, ransomware, and advanced persistent threat (APT) assaults [23].

Cloud inside an organization: The private cloud, also known as the internal cloud, is a kind of cloud that is used within an organization. This category emphasizes on a single user, group, or institution at the time of writing. Although the cost of private clouds is more than the cost of public clouds, they are more secure than public clouds. The fact that a private cloud is housed behind an enterprise's firewall allows users within the organization to access it via the company's intranet. Privatized clouds, in contrast to public cloud computing, are less secure since less money and experience is directed toward developing services and systems, much alone protecting data in the private cloud. Consequently, some components may become vulnerable, allowing hostile actors to conduct attacks against these vulnerable components by exploiting their weaknesses [23].

The community cloud assists various communities with common interests, such as missions, rules, security needs, and regulatory compliance difficulties, among other things. Institutions or a third party may manage it on-site or off-site, depending on the circumstances. The community cloud offers more robust privacy, security, and policy compliance protections than the standard cloud. The degree of security in a community cloud environment is determined by the quantity of security awareness in the community and the importance of security to the community's activities. The cloud storage of sensitive data from a government agency may endanger national security if the material is made available to the public, as has happened in the past. As a result, security measures should be included in cloud computing environments [24].

Hybrid cloud: Due to an institution's diverse needs, this kind of cloud deployment is required. It combines two or more models to deliver cloud-based computing services (public, private, or community). Enterprises may use private clouds to store sensitive data or apps in a secure environment while hosting nonsensitive data or applications in a public cloud environment. Because of the federation of clouds with a diverse set of incompatible security measures, cloud hybridization, on the other hand, generates a host of security challenges. A consequence is that attackers uncover vulnerabilities in one or more clouds to access the whole infrastructure [25]. This chapter aims to provide a new taxonomy for safe cloud architecture based on current literature, provide an in-depth review of various cloud infrastructure issues and solutions, and draw attention to the disadvantages of existing solutions.

The chapter aims to answer two critical questions:

i. What are the most well-known challenges in cloud computing architecture and proposed remedies at different levels of abstraction?
ii. What security dangers are associated with the cloud that might prevent its widespread use?

The research utilized academic digital resources, such as the ACM Digital Library and Arxiv, as well as international conferences. The extensive study between 2011 and 2020 involved searching large libraries using a combination of search phrases, including "Application Security" and "Network Security," to obtain reliable search results. Keywords and topics from abstracts were collected to emphasize the relevant contributions of the studies.

9.2 LITERATURE REVIEW

Over the previous decade, several survey studies have been published exploring the security risks of cloud computing. Regarding cloud security, the great majority of the information that has been evaluated has substantially contributed to managing these problems. One such study looked at the most often found cloud security flaws and discovered several. They also offered several additional solutions to security challenges that arise in cloud architecture, each of which was meant to be sensitive to the personal data of individual users. Data transfer through the cloud is subject to considerable security risks, according to research done by participants in this survey, who were provided practical advice on dealing with potential dangers during the survey. The study results included a taxonomy and survey of cloud services, which cloud infrastructure providers and revenue organized [8]. A service taxonomy was created, encompassing themes such as computers, networking, databases, storage, analytics, and machine learning, among other things, as well as additional topics. Regarding functionality, the computing, networking, and storage services provided by all cloud suppliers are high quality and commonly recognized as the backbone of the cloud computing architecture. According to a survey, cloud computing firms face several security issues. The cloud client, the CSP, and the owner of the data stored in the cloud all were involved in this process. An investigation of various communication and storage options in the crypto cloud was also conducted as part of the project. Researchers working on the causes and consequences of cyberattacks have access to the most up-to-date information [26–28].

Many data protection issues that may develop in a multi-tenant cloud computing system were examined, and solutions were provided in a study published by the researchers. While this poll focused more on data privacy than security, the prior survey was concerned with both concerns simultaneously. The released research gave a full definition of cloud computing and the many different levels of cloud architecture found in the cloud computing environment. The study included comparing three service models (including SaaS, PaaS, and IaaS) and three deployment methodologies as part of the overall research design (private, public, and community). It was determined that private and public clouds have information security needs; thus, the writers looked into it. A few of the most urgent difficulties and restrictions related to cloud computing regarding security were also covered during this session. According to a study published in the journal, one of the many vulnerabilities that often occur in cloud computing systems is the inability to recognize the flaws. In this research, the author's contribution consisted of categorizing different threats by the accessibility of cloud-based service resources. It was necessary to create this category in response to the extensive description and extent of the multiple dangers faced [29–31].

There are several concerns about the security of cloud computing infrastructure. Four critical levels of consideration should be considered while designing and executing cloud infrastructure security: the data level, the application level, the network level, and the host level (or the host itself) (or the physical location of the cloud infrastructure). First and foremost, security refers to the protection of programs using hardware and software resources to prevent others from gaining control. Among the most severe dangers at this level are distributed denial of service (DDoS) assaults on software programs, which are becoming more common. Second, network-level security concerns network protection via a virtual firewall, creating a demilitarized zone (DMZ), and data in transit protection procedures. Information about various firewalls should be monitored, collated, and preserved for future reference to achieve this goal [31]. Third, the degree of security refers to the protection offered for the host rather than for the virtual machine when a virtual server, hypervisor, or virtual machine is used in conjunction with another virtual machine. Obtaining information from system log files is required to know when and where applications have been recorded to make these determinations. When defending cloud infrastructure, looking at the primary CIA components at each level of the organizational hierarchy is critical. As cloud-based systems gain in popularity, the security dangers connected with their use are becoming better recognized. However, despite its many benefits, cloud computing is susceptible to various security risks and assaults. The cloud computing infrastructure is always under assault, and attackers constantly search for security flaws. The following parts discuss security issues that might arise at various levels of cloud architecture and how to solve them. Fourth, the data-level difficulties: At this level of complexity, topics such as data breaches, data loss, data segregation, virtualization, confidentiality, integrity, and availability all may be discovered [32].

In terms of application-specific options, there are a plethora of choices accessible. The authors have presented an ECC-based multi-server authentication approach that is specific to the MCC context and does not need any pairing on the part of the users. While saving time and money, this method also maintains the benefits of more expensive pairing systems, such as safe mutual authentication, anonymity, and scalability, without requiring extra resources. This is shown theoretically by the formal security model, which illustrates the method's robustness in practice. The authors used the Open Stack platform as a reference to develop several models for information and resource sharing among tenants in an IaaS cloud environment, which were then evaluated. A tenant is encouraged to interact with the IT resources of other tenants in a regulated manner by using the models provided. However, network access to virtual machines (VMs) must be regulated to prevent malicious software from moving data uncontrolled from the virtual machine.

According to the results, a unique access control architecture for cloud computing addresses cloud security and privacy challenges has been created. The notion of dynamic trustworthiness was the foundation for constructing the suggested system. An access control system based on dynamic trustworthiness is used to, among other things, minimize the probability of undesirable behavior and ensure that only authorized users have access to cloud resources. The results reveal that the system recognizes potentially dangerous actions to prevent unlawful access, which would improve cloud computing security and, as a consequence, raise user confidence in the system, according to the researchers. The authors presented a hybrid access control framework called iHAC, combining type enforcement and role-based access control with other techniques. As a result, the recommended architecture is universally applicable to IaaS cloud systems and allows for implementing highly flexible access control settings. An access control mechanism based on the virtual machine manager (VMM) was also created, which allows the VM's actions to be confined to the underlying resources at a finer level of detail. It has been shown in these researches that the implementation of the iHAC framework aids in selecting real-world access control choices while imposing an acceptable performance cost on the system under examination [33].

Another research showed that dynamic access control may be utilized to handle the many security threats in a cloud setting. Through this technique, it is feasible to safeguard cloud data by considering the interrelationship between the requestor, the data being sought, and the action that will be taken on that data. The demands of the user were taken into account as well while offering dynamic access control. A first attempt at putting the anticipated method into action resulted in the outcome. The authors provided network-level solutions such as SNORT, an intrusion detection system for cloud computing, under the Network-Level Solutions section as a network-level solution to prevent DoS and DDoS assaults. Such an attack floods the server with unnecessary packets, rendering it unusable for genuine users [33].

To recognize and prevent DDoS assaults, the suggested system uses specific criteria set in advance of implementation. The authors outlined a strategy along the same lines and showed a mechanism for recognizing and filtering diverse DDoS assaults in cloud-based systems. When constructing this strategy, it is vital to use both the GARCH model and an artificial neural network (ANN) to get the best results. When the actual value of variances is compared to a specified value of variances, Garch calculates the value of variances and finds any probable anomalies in real-world traffic. After values that are less than a certain threshold are eliminated, the ANN is used to categorize traffic into two categories: normal and anomalous. Regular traffic is defined as traffic that is less than a certain threshold.

Following the publication of a new article, users may randomly encrypt and push data blocks in a peer-to-peer network based on blockchain by using a technique detailed in the study. In certain circumstances, several data centers and users in a distributed cloud might complicate the placement of file block copies, leading to performance issues. As a result, it seems that the blockchain technique is the most favorable regarding file security and network transmission time, respectively. Another research presented dynamic proof to aid in public audibility in the case of data corruption by combining irretrievability methods with communication-efficient recovery strategies in the event of data corruption. The suggested technique might be used in storage to decrease the effect of modifications on data stored in a different place from the one being modified. Therefore, any effort to update will have only a minor influence on the actual codeword symbols. In a server failure, a dependable data reformatting technique may be used to restore data integrity. Using the domain name system as a springboard, the authors developed a thorough list of DNS assaults. Firewall use is the most common approach to DNS strategy, and this is considered one of the best practices in establishing DNS servers, according to the authors of this chapter. The dynamic DNS firewall provides an additional layer of security and carefully created signatures [33].

OpenPipe SaaS was invented by researchers and used by the industry. A hybrid control mode was used to implement it, with the top level being a software-defined networks (SDN) controller and the bottom level being local controllers, using the hybrid control mode. SDN's separation of the control plane from the data plane was expected to provide several advantages, including network virtualization and programmability. OpenPipe was shown in a laboratory environment. Certificates, higher level–based authentication, and other encryption-based measures have been demonstrated to be successful in protecting cloud computing environments against unauthorized access, according to the findings of the study (e.g., symmetric and asymmetric key algorithms). They suggested a Bayesian network-based weighted attack route modeling approach to model attack pathways to understand better how they operate. Moreover, they presented an enhanced technique for determining the most direct and least expensive attack vector from many sources, using crucial nodes and critical edges. Apart from choosing the most direct path between two points of interest, the algorithm also dismantles links between routes of equal significance. There are several alternatives to data, but one of the most essential is focusing intensely on data security and privacy as we migrate from traditional computer models to the Internet-based cloud computing paradigm. It is possible that data loss or leakage may have a substantial impact on a company's bottom line and will cause customers to lose confidence in the company's product or service. Recent research has thoroughly examined the auditing procedure in a cloud computing environment. When it comes to data auditing, it is necessary to check for a range of characteristics, including confidentiality, integrity, remanence, provenance, and lineage, among other things. Following the research, each of these concerns has a set of basic procedures that, except data remanence, which is still a hot topic in public cloud services, may be able to meet the data auditing requirements of cloud service users.

A third-party auditor is responsible for ensuring that client data stored on a cloud storage server is correct and complete. This was the theme on which the authors focused. It has been discovered that a technique for dynamic data updates has been devised utilizing an improved Chameleon Authentication Tree. By demonstrating that their enhanced auditing protocol is immune to assaults such as replay, replace, and forge, the researchers could further illustrate the protocol's security. A categorization method based on a range of criteria has been devised due to the research outcomes. The parameters were selected by examining various aspects of the problem. It is meant to give varied degrees of protection depending on the kind of material utilized and the degree of accessibility. In accordance with the authors' findings, data security may be offered at various degrees of protection, depending on the amount of protection required. Security precautions for storage may be implemented depending on the dataset that has been classed as dimensions in the database and can be enforced accordingly.

Secure data classification is the name the authors give to a cloud computing strategy based on safe data categorization and classification. Using TLS, AES, and SHA cryptographic methods, which are chosen depending on the kind of classified data, minimizes the total time necessary to protect data. The inquiry results reveal that the proposed model has been thoroughly tested and is reliable and effective. They devised a privacy-preserving paradigm for outsourced categorization (POCC), which they used as a case study for cloud computing. When training a POCC model utilizing encrypted data distributed across many sources, the evaluator may be confident that the model's classification accuracy and dependability will not suffer. The authors used Gentry's approach to create the world's first entirely homomorphism encryption system, which they used to protect sensitive data throughout development [34–37].

9.3 OPEN CHALLENGES

The National Science Foundation says cloud computing research is still in its infancy despite companies and sectors' widespread use of the technology. Most cloud infrastructure gaps have not yet been closed, and new issues are always on the horizon. The following sections give an overview of the most critical open issues that require further investigation. *Hypervisor security:* The security of cloud computing is jeopardized if the hypervisor is compromised. It has the potential to do serious harm to the whole network. Because of the cloud's dynamic nature, traditional detection and prevention methods are no longer helpful. These technologies are critical for distinguishing between normal and aberrant cloud computing behavior. In addition, any recommended treatment must be executed as quickly as feasible to prevent damaging the cloud infrastructure or interfering with routine operations [38].

A third-party auditing firm: Data loss and erasure due to hardware–software failures and/or human mistakes have been highlighted as worries regarding the integrity of cloud-based information storage systems expand in popularity and reach. A third-party auditor independent of the firm should provide expert integrity verification services. Public cloud information auditing requires that a client's private information not be given to any public verifier throughout the process. Consequently, a new privacy-related main concern has emerged: the risk of third-party auditors accessing personal data. Researchers are always looking for ways to keep cloud storage safe and secure. When a security breach occurs, the system must function normally again. It is described as the extent to which data, software, and hardware are made accessible to authorized users in answer to their requests. The framework can perform duties at any hour of the day or night that is considered system availability. Three of the cloud environment's most confusing challenges have been data protection, availability, and security. When data is destroyed, reformatted, or redistributed to a new user, it is known as data remanence. As a result, the privacy of erased files is at risk. Computer forensics and other methods may be used to identify data remanence. Data recovery software may also recover data accidentally deleted from a computer. Cloud providers have put little to no effort into addressing the issue of data remanence, even though it is one of the most pressing issues. It was falsely claimed that IaaS's suggested security measures would secure the network; however, this was not the case. Some assaults are insurmountable by a standard firewall, but not all. Cloud computing is getting riskier because of the number of assaults recorded due to the DNS hit. Because of the lack of investigation into the problem of reusing IP addresses, serious data and system breaches have occurred, endangering customers' privacy and data security. IaaS security can't be strengthened by traditional access control and identity management systems because of the prominent cloud-specific properties. Computer security in the modern world requires cutting-edge technology like blockchain and computational intelligence. The great majority of authentication methods are both time-consuming and challenging to implement. Compared to more conventional techniques, existing research used simulation to evaluate their ideas with a small quantity of data, rudimentary resources, and few users. Ultimately, though, the cloud is a complex system with a large number of users and a variety of other variables. A more significant effort should be made to design approaches that address these limitations. On the other hand, the CSP does not offer a platform that enables simultaneous usage of different user interfaces for authentication [39].

9.4 RECOMMENDATIONS

The following are some recommendations based on the present challenges:

Context-aware solutions are necessary to avoid potential harm to the cloud infrastructure. These solutions must identify new and changing attack patterns and react quickly to prevent any potential harm to the cloud infrastructure. Consider both the client's preferences and the extent of the client's understanding of security while creating these solutions.

The following aspects should be considered while developing third-party auditing solutions for third parties: performing third-party audits without having access to the examined data ensures that the privacy of the data is always protected. It is advised that the data be divided and encrypted in a cloud storage system to guarantee that it is always kept hidden.

Detecting whether or not the stored data has been tampered with and alerting the user of the results is performed at the client's request. Data is available upon request. How data is stored affects how easily accessible information is to consumers. As indicated in the following points, a variety of ways may be used to assure data availability, and these strategies may be the topic of future study in the area of cloud infrastructure security: to ensure data security, backups must be kept on an individual user basis, or in a widely scattered network environment. This means that if the storage component fails or degrades, the user will not be forced to delete all his or her data. Frequent updating of backups is required so that the user may always access the most up-to-date versions of the data stored on the system. According to the company, data loss prevention (DLP) solutions assist in preventing data breaches and reducing physical damage to data center equipment and infrastructure. Many technologies rely on third-party cloud-based secure storage to keep their data safe and secure and to prevent it from being lost or stolen. Many data loss protection programs include monitoring, threat blocking, and forensic analysis. Object storage uses sophisticated erasure coding techniques to assure data availability and integrity. It describes combining data with parity data before breaking and distributing it throughout a storage area using "erasure-coded" data.

Residual data: As seen in the following table, several solutions may be utilized to remove or minimize residual data. In certain circles, sterilization is referred to as "purging," which relates to eliminating sensitive data from a storage system to prevent it from being recovered via a recognized method or technology. Encryption is a very effective means of keeping your data safe and secure.

Network security measures: Some ideas for securing your network against cyberattacks are the following. Secure communication protocols such as HTTPS and TLS (transport layer security) must be used to protect cloud internal communications. There are several ways to identify and prevent hazardous network intrusions using HTTP request anomaly detection. You may quickly get your hands on these solutions.

Access control and identity management are two critical parts of operating a successful company. The following security issues should be included in any future research on access control and identity management. Access key authentication should be the only way to access the cloud. All assets and business systems in the cloud should be organized and classified to ensure SOM in the cloud includes the following:

a. Situation awareness
b. Safe operation and maintenance, which incorporates unified ID authentication
c. Unified account management
d. SSO, or single sign-on, which is already standard practice in many cloud environments, including AWS. One way to utilize SSO is with blockchain-based self-sovereign identity management systems. This update will make a standard and private means of storing and maintaining credentials available to customers.

Security: Because of the complexity of the resources, the huge number of users, and the varied nature of the cloud, it is vital to utilize time-saving authentication mechanisms. Also required are authentication methods for various user interface authentications. According to some experts, blockchain technology may be leveraged to produce more secure authentication techniques shortly.

9.5 CONCLUSIONS

The three categories of cloud computing services are available: SaaS, PaaS, and IaaS, respectively. Customers and service providers have equal levels of responsibility for the upkeep and protection of the environment. Different kinds of cloud computing environments are referred to by various names, including public, private, communal, and hybrid clouds. Studies have been conducted to investigate the security flaws present in cloud computing systems, concentrating on the data, application, network, and host levels. ECC-based multi-server authentication, Open Stack platform models, iHAC access control architecture, network-level solutions, dynamic proof techniques, and Open Pipe SaaS are some innovations researchers have produced. Data security and privacy are becoming more critical as more and more enterprises transition to the Internet-based cloud computing paradigm. A few loopholes in the infrastructure need to be filled, such as hypervisor security, third-party audits, and data protection. In the present level of safety and protection, sophisticated technologies like blockchain and computational intelligence are required. Solutions that are aware of their context, audits by a third party that does not need access to the data, and network security measures such as HTTPS, TLS, and HTTP request anomaly detection are required.

REFERENCES

1. V. Bs, M. Arvindhan, and S. Kalimuthu, "The crucial function that clouds access security brokers play in ensuring the safety of cloud computing," 2023 16th International Conference on Security of Information and Networks (SIN), Jaipur, India, 2023, pp. 1–5. doi: 10.1109/SIN60469.2023.10475014
2. A. Joshi, A. Raturi, S. Kumar, A. Dumka, and D. P. Singh, "Improved security and privacy in cloud data security and privacy: measures and attacks," 2022 International Conference on Fourth Industrial Revolution Based Technology and Practices (ICFIRTP), Uttarakhand, India, 2022, pp. 230–233. doi: 10.1109/ICFIRTP56122.2022.10063186
3. R. Deepika and D. Kumar, "Security enabled framework to access information in cloud environment," 2022 International Conference on Machine Learning, Big Data, Cloud and Parallel Computing (COM-IT-CON), Faridabad, India, 2022, pp. 578–582. doi: 10.1109/COM-IT-CON54601.2022.9850906.
4. O. Mejri, D. Yang, and I. Doh, "Cloud security issues and log-based proactive strategy," 2021 23rd International Conference on Advanced Communication Technology (ICACT), PyeongChang, Korea (South), 2021, pp. 392–397. doi: 10.23919/ICACT51234.2021.9370392
5. O. Mejri, D. Yang, and I. Doh, "Cloud security issues and log-based proactive strategy," 2022 24th International Conference on Advanced Communication Technology (ICACT), PyeongChang, Korea, 2022, pp. 392–397. doi: 10.23919/ICACT53585.2022.9728822
6. T. Eltaeib and N. Islam, "Taxonomy of challenges in cloud security," 2021 8th IEEE International Conference on Cyber Security and Cloud Computing (CSCloud)/2021 7th IEEE International Conference on Edge Computing and Scalable Cloud (EdgeCom), Washington, DC, USA, 2021, pp. 42–46. doi: 10.1109/CSCloud-EdgeCom52276.2021.00018
7. H. Kimm and J. Ortiz, "Multilevel security embedded information retrieval and tracking on cloud environments," 2021 IEEE Cloud Summit (Cloud Summit), Hempstead, NY, USA, 2021, pp. 25–28. doi: 10.1109/IEEECloudSummit52029.2021.00012

8. P. Asmitha, C. Rupa, S. Nikitha, J. Hemalatha, and A. K. Sahu, Improved multiview biometric object detection for anti spoofing frauds. Multimedia Tools and Applications, pp. 1–17 (2024). https://doi.org/10.1007/s11042-024-18458-8
9. A. K. Sahu, K. Umachandran, V. D. Biradar, O. Comfort, V. Sri Vigna Hema, F. Odimegwu, and M. A. Saifullah, A study on content tampering in multimedia watermarking. SN Computer Science, 4(3), p. 222 (2023).
10. M. A. Helmiawan, I. Fadil, Y. Sofiyan, and E. Firmansyah, "Security model using intrusion detection system on cloud computing security management," 2021 9th International Conference on Cyber and IT Service Management (CITSM), Bengkulu, Indonesia, 2021, pp. 1–5. doi: 10.1109/CITSM52892.2021.9588810
11. S. Li, F. Dang, Y. Yang, H. Liu, and Y. Song, "Research on computer network security protection system based on level protection in cloud computing environment," 2021 IEEE International Conference on Advances in Electrical Engineering and Computer Applications (AEECA), Dalian, China, 2021, pp. 428–431. doi: 10.1109/AEECA52519.2021.9574216
12. S. Aparajit, R. Shah, R. Chopdekar, and R. Patil, "Data protection: The cloud security perspective," 2022 3rd International Conference for Emerging Technology (INCET), Belgaum, India, 2022, pp. 1–5. doi: 10.1109/INCET54531.2022.9825151
13. M. V. Reddy, P. S. Charan, D. Devisaran, R. Shankar, and P. M. Ashok Kumar, "A systematic approach towards security concerns in cloud," 2023 Second International Conference on Electronics and Renewable Systems (ICEARS), Tuticorin, India, 2023, pp. 838–843. doi: 10.1109/ICEARS56392.2023.10085437
14. S. R. Jena, V. Vijayaraja, and A. K. Sahu, Performance evaluation of energy efficient power models for digital cloud. Indian Journal of Science and Technology, 9(48), p. 222 (2016).
15. N. Tutubala and T. E. Mathonsi, "A hybrid framework to improve data security in cloud computing," 2021 Big Data, Knowledge and Control Systems Engineering (BdKCSE), Sofia, Bulgaria, 2021, pp. 1–5. doi: 10.1109/BdKCSE53180.2021.9627294
16. F. Dang, L. Yan, and Y. Yang, "Research on intelligent centralized system based on security architecture of computer cloud security protection," 2023 IEEE 3rd International Conference on Electronic Technology, Communication and Information (ICETCI), Changchun, China, 2023, pp. 1281–1285. doi: 10.1109/ICETCI57876.2023.10176977
17. C. Choudhary, N. Vyas, and U. Kumar Lilhore, "Cloud security: Challenges and strategies for ensuring data protection," 2023 3rd International Conference on Technological Advancements in Computational Sciences (ICTACS), Tashkent, Uzbekistan, 2023, pp. 669–673. doi: 10.1109/ICTACS59847.2023.10390302
18. V. Bs, M. Arvindhan, B. B. Kannan, and S. Kalimuthu, "The crucial function that clouds access security brokers play in ensuring the safety of cloud computing," 2023 International Conference on Communication, Security and Artificial Intelligence (ICCSAI), Greater Noida, India, 2023, pp. 98–102. doi: 10.1109/ICCSAI59793.2023.10420940
19. K. R. Raghunandan, R. Dodmane, K. Bhavya, N. K. Rao, and A. K. Sahu, "Chaotic-map based encryption for 3D point and 3D mesh fog data in edge computing." IEEE Access, 11, pp. 3545–3554 (2022).
20. K. Kanagasabapathi, K. Mahajan, S. Ahamad, E. Soumya, and S. Barthwal, "AI-enhanced multi-cloud security management: Ensuring robust cybersecurity in hybrid cloud environments," 2023 International Conference on Innovative Computing, Intelligent Communication and Smart Electrical Systems (ICSES), Chennai, India, 2023, pp. 1–6. doi: 10.1109/ICSES60034.2023.10465550

21. A. Syed, K. Purushotham, and G. Shidaganti, "Cloud storage security risks, practices and measures: A review," 2020 IEEE International Conference for Innovation in Technology (INOCON), Bengaluru, India, 2020, pp. 1–4. doi: 10.1109/INOCON50539.2020.9298281
22. S. Gahane and P. Verma, "The research study on identification of threats and security techniques in cloud environment," 2023 1st DMIHER International Conference on Artificial Intelligence in Education and Industry 4.0 (IDICAIEI), Wardha, India, 2023, pp. 1–6. doi: 10.1109/IDICAIEI58380.2023.10407003
23. S. An, A. Leung, J. B. Hong, T. Eom, and J. S. Park, Toward automated security analysis and enforcement for cloud computing using graphical models for security. IEEE Access, 10, pp. 75117–75134 (2022). doi: 10.1109/ACCESS.2022.3190545
24. J. Chavan, R. Patil, S. Patil, V. Gutte, and S. Karande, "A survey on security threats in cloud computing service models," 2022 6th International Conference on Intelligent Computing and Control Systems (ICICCS), Madurai, India, 2022, pp. 574–580. doi: 10.1109/ICICCS53718.2022.9788148
25. L. M. Brumă, "Cloud security audit: Issues and challenges," 2021 16th International Conference on Computer Science & Education (ICCSE), Lancaster, UK, 2021, pp. 263–266. doi: 10.1109/ICCSE51940.2021.9569654
26. Z. Tang, "A preliminary study on data security technology in big data cloud computing environment," 2020 International Conference on Big Data & Artificial Intelligence & Software Engineering (ICBASE), Bangkok, Thailand, 2020, pp. 27–30. doi: 10.1109/ICBASE51474.2020.00013
27. R. K. Nema, A. K. Saxena, and R. Srivastava, "Survey of the security algorithms over cloud environment to protect information," 2022 10th International Conference on Emerging Trends in Engineering and Technology: Signal and Information Processing (ICETET-SIP-22), Nagpur, India, 2022, pp. 1–6. doi: 10.1109/ICETET-SIP-2254415.2022.9791643
28. D. Li, L. Yan, Y. Song, S. Li, and H. Liang, "Network computer security and protection measures based on information security risk in cloud computing environment," 2021 IEEE International Conference on Advances in Electrical Engineering and Computer Applications (AEECA), Dalian, China, 2021, pp. 424–427. doi: 10.1109/AEECA52519.2021.9574194
29. R. Kumar and M. P. S. Bhatia, "A systematic review of the security in cloud computing: Data integrity, confidentiality and availability," 2020 IEEE International Conference on Computing, Power and Communication Technologies (GUCON), Greater Noida, India, 2020, pp. 334–337. doi: 10.1109/GUCON48875.2020.9231255
30. Z. Zou, "Research on user information security based on cloud computing," 2023 IEEE 7th Information Technology and Mechatronics Engineering Conference (ITOEC), Chongqing, China, 2023, pp. 35–39. doi: 10.1109/ITOEC57671.2023.10291704
31. A. K. Sahu and G. Swain, "A novel multi stego-image based data hiding method for gray scale image." Pertanika Journal of Science & Technology, 27(2), pp. 753–768 (2019).
32. E. M. Toth and M. M. Chowdhury, "Honeynets and cloud security," 2022 IEEE World AI IoT Congress (AIIoT), Seattle, WA, USA, 2022, pp. 270–275. doi: 10.1109/AIIoT54504.2022.9817263.
33. G. Feng, Q. Huang, Z. Deng, H. Zou, and J. Zhang, "Research on cloud security construction of power grid in smart era," 2022 IEEE 2nd International Conference on Data Science and Computer Application (ICDSCA), Dalian, China, 2022, pp. 976–980. doi: 10.1109/ICDSCA56264.2022.9987863

34. A. Kaur, A. Dhiman, and M. Singh, "Comprehensive review: Security challenges and countermeasures for big data security in cloud computing," 2023 7th International Conference on Electronics, Materials Engineering & Nano-Technology (IEMENTech), Kolkata, India, 2023, pp. 1–6. doi: 10.1109/IEMENTech60402.2023.10423449
35. S. Deb, A. Das, and N. Kar, An applied image cryptosystem on Moore's automaton operating on δ(qk)/𝔽2. ACM Transactions on Multimedia Computing, Communications and Applications, 20(2), pp. 1–20 (2023).
36. S. Deb, S. Pal, and B. Bhuyan, "NMRMG: Nonlinear multiple-recursive matrix generator design approaches and its randomness analysis." Wireless Personal Communications, 125(1), pp. 577–597 (2022).
37. K. S. Roy, S. Deb, and H. K. Kalita, A novel hybrid authentication protocol utilizing lattice-based cryptography for IoT devices in fog networks. Digital Communications and Networks, 10(1) (2022).
38. E. R. Kumar, S. S. S. Reddy and M. B. Reddy, "A multi-stage cloud security for cloud data using amalgamate data security," 2023 International Conference for Advancement in Technology (ICONAT), Goa, India, 2023, pp. 1–5. doi: 10.1109/ICONAT57137.2023.10080583
39. S. Mishra, M. Kumar, N. Singh, and S. Dwivedi, "A survey on AWS cloud computing security challenges & solutions," 2022 6th International Conference on Intelligent Computing and Control Systems (ICICCS), Madurai, India, 2022, pp. 614–617. doi: 10.1109/ICICCS53718.2022.9788254